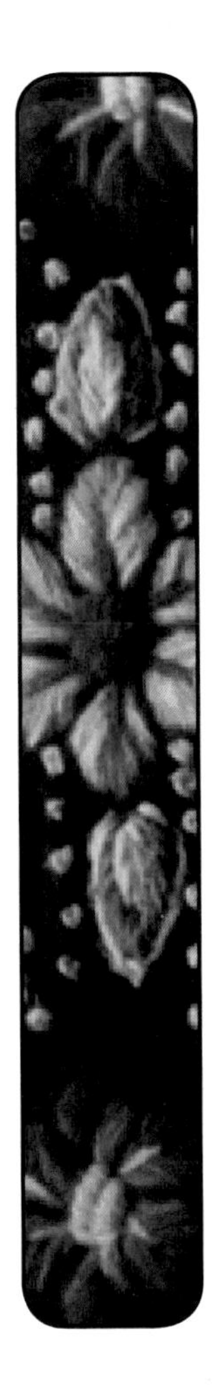

PERU

GATEWAYS TO THE ANDES

(The Alternative Travel Guide)

John Lane

November 2017
(First Edition)

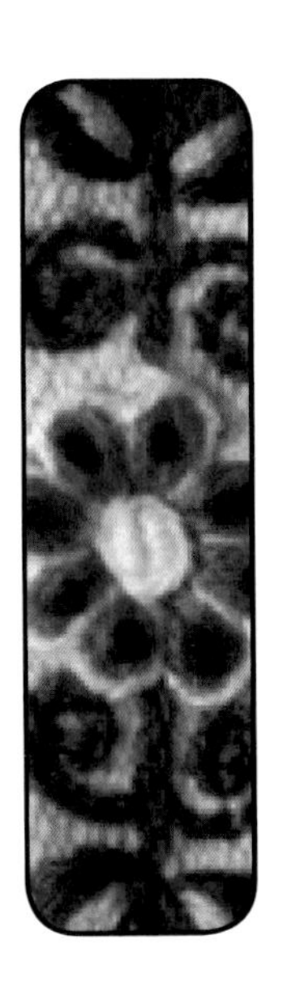

New World Books
London

For
Bellita and Josito
my hugely talented and inspirational guides
on the path less travelled

Peru - Gateways to the Andes
(The Alternative Travel Guide)

First Published November 2017

New World Books

(in collaboration with Jose Navarro and Bella Lane)

A CIP catalogue record for this title is available from the British Library

ISBN 978-0-9934021-3-5

Printed and bound in Great Britain by Witley Press, Hunstanton, Norfolk
www.witleypress.co.uk

Photography by author and collaboraters, except as attributed and by kind courtesy

Contents

Presidential Palace - Lima.

Historic Jirón Trujillo, just over the Puente Piedra across the Rimac.

Mercado Surquillo.

Parque Reducto - defensive redoubt captured by the Chileans January 1881; now with a museum dedicated to Mariscal Cáceres.

Locomotive 'President Leguía' - Parque Reducto, Lima. Recovered for posterity from the Santa Ana stretch of the Cusco railway.

Seller of quails eggs - five for one sol for your breakfast.

Prologue
Lima

"Gastronomic capital of the Americas"

Arriving in Peru by Air

Entry is through Jorge Chavez International Airport, Lima . There are plans to extend the runaway at Cusco to accommodate overseas flights in the future, and in the past there have been international flights directly into Iquitos from Panama City and elsewhere, and similar services (for example, from Bogota and from the States) are again being considered. But for the moment, Lima is the sole point of access to the country by air. Airport taxes for external and domestic flights are now invariably included in the ticket price at the point of purchase. Taxis to downtown Lima etc (e.g. Miraflores) arranged within the airport will cost S/.60 and upwards; outside taxis may be had for S/.40 or less.

The flight permutations are many: British Airways fly direct from Gatwick, London, albeit not all year round; KLM fly from Amsterdam; Iberia from Madrid and Air France from Paris. Alternative routes lie through North America: Air Canada via Toronto; US airlines via Houston, Newark, Atlanta and Miami. Travelling via Mexico City is another colourful option, or even through La Paz, Rio de Janeiro, Buenos Aires or Quito. Avianca 787 Dreamliners fly from London Heathrow through Bogota to Lima, whilst from Australia the LAN/ Quantas A380's run from Sydney to Santiago, usually via New Zealand. The range of modern airliners is ever-increasing; gone are the days of re-fuelling in Aruba or Bonaire in the Caribbean.

Lima – Potted Profile

The city was founded in 1535 by Conquistador Francisco Pizarro. Present day population is of the order of 10 million – one third of the total population of Peru. The climate is humid with summer temperatures (December to April) around 30°C; the remainder of the year is marked by the presence of the *garua* mist (hence '*Lima la Gris*') with daily averages of 15°C. Annual rainfall is negligible. Lima telephone code: 01 (or with country code +51-1-etc). In 1746 the city suffered an earthquake estimated at 9.0 on the Richter scale; the earthquake which devastated the city in 1940 measured 8.2 on the Richter scale. The University of San Marcos in Lima is the oldest higher learning institution in the New World. In 2019 the city will host the Pan American Games. Lima is the second largest city in South America – and one of the most vibrant to visit.

Accommodation in Lima

There are countless hotels from which to choose with the aid of the internet. Recently extended and renovated Hotel Antigua Miraflores at 350 Av. Grau is something of a collector's gem, full of artefacts, plus a beautiful little garden (tel: 201-2060). Moreover, it is an easy walk to the offices of Peruvian Airlines, LAN Peru, Star Peru and LC (*Luis Carlos*) Peru, conveniently clustered adjacent to each other on Av. Jose Pardo (just 3-5 blocks down from Ovalo Miraflores, reference point department store *Saga Falabella*). The smallest of these, LC Peru, is worthy of special attention in so much that it specialises particularly in serving some of the more remote Andean destinations, such as Ayacucho, Andahuaylas, Cajamarca, Huánuco, Huaraz, Jauja and Tingo María.

Returning to the subject of accommodation, for a treat consider staying at centrally placed Hotel Maury (tel: 428-8188) where half a century or so ago 'the Peruvian winner of the Classic South American sprint at Lima's *Hipódromo de Monterrico* racecourse was historically ridden to the counter of the eponymous Dardanus Bar to partake of a silver pail of pisco sour', or at equally historic Gran Hotel Bolivar (tel: 619-7171) on Plaza San Martin, inaugurated in 1924 marking the centenary of the Battle of Ayacucho and coincidentally claiming to serve the best Pisco Sours anywhere. Or indeed spend time at the outstanding Belmond Miraflores Park (tel: 610.4000) adjacent to the Larcomar mall overlooking the Pacific. More modestly try welcoming Hotel San Antonio Abad at Ramon Ribeyro 301 (tel:447.6766), handy for Parque Reducto, Mercado Surquillo and Gimnasio Mario No.1 at 156 V.Prolong Ricardo Palma but rather further to Fitness at Miguel Dasso (now that Gold's on Av.Benavides has closed).

Finally, and when your thoughts turn to Cusco and Machu Picchu and the possibility of other visits to alternative destinations such as Arequipa and the Colca Canyon, Ayacucho and Andahuaylas, and even the magical ruins of the Chachapoya dynasty at Kuelap, then for convenience and early morning flights, the Ramada Costa del Sol Hotel at Lima airport is ideal (tel: 711-2000).

Lima – Where, What & How

By-passing Lima would be an error, the city is packed with history, architecture, museums, restaurants, markets and malls, sights and general interest, all well documented. Programme at least two full days, and better still, a full week to do justice to the capital if you have time. Taxis for all your excursions are of course readily available, but in view of the perpetual heavy traffic and resultant pollution do not overlook the **Metropolitano** mass transit bus service introduced in 2010, which is both rapid and economic, if rather squashed (buy a travel card at any entry point), plus of course the regular buses which serve all areas.

In addition to the well advertised standard activities, which include of course the Plaza de Armas and surroundings, the daily guard-changing ceremony in front of the Presidential Palace at 1145-1200 (and at 1100 on Sundays), the Cathedral, the Archbishop's Palace, and all the many museums of interest, before you leave Lima be sure to visit the *Parque de Las Leyendas* in San Miguel, and the *Museo del Banco Central de Reserva del Perú* (carry identification, such as a passport photocopy) in the centre of Lima (Jr.Ucayali), the Larco Mar cliff top mall with its vista of Horseshoe Bay and the coastline (and La Bonbonniere restaurant), *Museo* Pedro D.Osma in Barranco, the naval museum at Callao (www.museonaval.com.pe), and the fascinating Royal Felipe Fortress also in the port of Callao. Summer beaches may be found at Barranco and Chorillos, including the Fish Market at the latter. Enjoy also Parque Kennedy, cutting through from Ovalo Miraflores to seek treasure in the antique and silver shops of Av. La Paz, stopping off at restaurant *Rimcon Chami* en route for lunch. And two famous restaurants to keep in mind for dinner are the Rosa Nautica (Lima Bay – tel: 445-0149) and La Costa Verde (Barranquito Beach – tel:247-1244).

Twenty years ago *Museo de la Nación* housed in the hugely impressive *Banco de la Nación* building was formerly an excellent starter-venue for seeing artefacts and displays on the rich cultural heritage of Peru. For reasons unclear this fine capacious edifice is nowadays an empty shell; not worth the trek out to Javier Prado Este 2465. However, upstairs there is a powerful and moving exhibition of *La Violencia 1980-2000* (Tues-Sun 0900-1700 except Public Holidays).

For those with an eye for curiosities, visit the *Puente de Piedra* dating from 1610 across the apologetic River Rimac behind the *Palacio de Gobierno* and the Desamparados railway station. There is a second so-called *Puente Piedra* on the main highway some distance out of Lima to the north, which is also known as the 'Bridge of Eggs' because when it was built in 1608 the cement for the masonry was mixed using the whites of 10,000 sea bird eggs, resulting in a highly binding recipe.

In January 2012 the first line of the new (and very long awaited) cross-city **mass transit urban rail network** was opened in Lima (the financial irregularities which caused years of delay having been set aside). Feasibility studies were commissioned as long ago as 1965 when trams were discontinued in the city; President Alan Garcia picked up the project in his first term of office and notwithstanding opposition managed to inaugurate two kilometres of track in April 1990 before standing down in July 1990. His successor (President Fujimori) then suspended work; a second presidential term for Alan Garcia (2006-2011) enabled him re-activate his dream.

On account of the elevation of the line at roof top level (rather than underground), this provides an excellent panoramic overview of the metropolis. And if you do indulge in a sample journey, consider jumping off at the Presbitero Maestro stop to view the historic and astonishing Lima General Cemetery, oldest monumental burial ground in the Americas (but be sure to visit in a group – the area is a hotspot for assaults). From one zany locale to another, try the *Polvos Azules* market at Paseo de la Republica – Antonia Raymondi – José Galvez: tops for all electronic items, cameras, smart phones, videos, and so on. The wholesale fun market for everything is *Mercado Central,* Av.Abancay/ Jr.Cusco. And the often overlooked venue for a colourful display of Peru's rich heritage of folk dancing before you head off into the mountains is *Asociacíon Cultural Brisas del Titicaca* (Jr.Heroes de Tarapaca #168, Alt.Cdra 1 Av.Brasil) where spectacular *Noches de Folclor* shows are held six evenings a week (book online early to reserve).

Altitude Sickness (*Soroche*).

An important consideration and conveniently included here at the outset of this guide to travelling in the High Andes. The elevation at which **altitude sickness** (*soroche*) manifests itself is generally considered to be 2,400m/8,000ft, which effectively means more or less just about anywhere in the mountains of Peru. *Soroche* or **AMS (Acute Mountain Sickness)** is no respecter of persons and is indiscriminate – some will suffer, some will be less affected. There is no pre-determinable profile to indicate likelihood of susceptibility, except that usually those who have previously been to moderate altitude without ill-effect will continue to enjoy a degree of immunity. There is a wealth of further professional medical information on **AMS** on the Internet, including perspective on the dangers of pulmonary and cerebral oedema, and preventative medications which may be professionally prescribed..

The effect on individuals of mild altitude sickness differs widely and is unpredictable, ranging through loss of appetite, severe headaches, hyperventilation, nausea, fatigue, insomnia, disorientation and dizziness – all stemming from the shortage of oxygen (also resulting in breathlessness after exertion). Above 3,048m/10,000ft ascent acclimatisation of 305m/1,000ft per day is recommended. This is generally well known to mountain climbers, but for tourists flying in or travelling by road or train this expedient is not necessarily readily available, although mountain passes are exactly that and are soon left behind.

'Sorojchi Pills' (sic) are on sale at pharmacies in Peru. Alternative aids/ placebos are available, including *coca* leaves (indeed a constituent of cocaine but not to be tarred by the same brush). The leaves of this plant – held to be sacred in the mountains – are widely chewed by *los Andinos*, or drunk as tea, and are equally advocated for visitors, at very least serving the recommendation of constant hydration. However, in severe cases of altitude sickness with persistent and prevailing symptoms the only way to alleviate suffering (aside from acclimatisation) is to descend, seeking medical help as necessary.

A reminder of the altitudes of some of the towns and cities and other points mentioned in this guide is as follows:

Abancay	**2377m/7,799ft**	**Andahuaylas**	**2926m/9,600ft**
Arequipa	**2350m/7,709ft**	**Ayacucho**	**2750m/8,250ft**
Cajamarca	**2750m/9,022ft**	**Chachapoyas**	**2335m/7,661ft**
Chivay	**3600m/11,811ft**	**Chonta Pass**	**4853m/15,922ft**
Cusco	**3399m/11,152ft**	**Huancayo**	**3271m/10,737ft**
Huancavelica	**3660m/12,008ft**	**Huánuco**	**1880m/6,168ft**
Huascarán Mt	**6768m/22,205ft**	**Jaén**	**729m/2,392ft**
Jauja	**3400m/11,155ft**	**Kuelap**	**3000m/9,843ft**
Leymebamba	**2203m/7,228ft**	**Machu Picchu**	**2430m/7,972ft**
Moyobamba	**860m/2,822ft**	**Ticlio Pass**	**4800m/15,748ft**

Plaza de Armas, Arequipa

Centro Historico, Arequipa

Platos tipicos

Contrasts in Arequipa

La Recoleta Monastery

Gustave Eiffel's iron bridge over el río Chili, dating from 1864

Section I
The Southern Highlands

"When you come to a fork in the road, take it"

Arequipa

Getting to Arequipa. Reached by air from Lima in 70 minutes (Peruvian Airlines, Star Peru and LAN all fly daily); factor in airport down time, plus to/from, say five hours. By road the 1,000km journey varies between 16 and 19 hours; the lead bus company is Cruz del Sur leaving from Javier Prado Este 1109; needless to say, there are many other operators, although Ormeño no longer run this route.

For this marathon inevitably half your trip will be made in the dark, but as always the best experience is top deck, front row (*panoramico*), for which the single fare is S/.110.00 (reserve in advance). Departure on the 1430 service will see you beyond Ica by dusk, straight down the *Panamericana del Sur* up to that point, after which you leave the dual carriageway and turn inwards and twist upwards (*Gravol* travel pills suggested) through the southern highlands of the Andes to reach the bus depot in Arequipa (Terminals Terrestre & Terrapuerto) conveniently between 8 and 9 a.m. (earlier departures from Lima will deposit you in the terminal in the small hours).

Potted Profile. Arequipa is the second largest city of Peru, with one million inhabitants. It lies in a valley overlooked by the snow-capped cone of volcano Misti (5822m/19,101ft) together with the extinct volcanic groups of Pichu Pichu and Chachani. The average height above sea level is 2350m (7709ft), just below the elevation at which **altitude sickness** (*soroche*) can occur, as already mentioned in the **Prologue** (p.4); susceptibility is unpredictable. *Coca* leaves chewed or drunk as tea are recommended. In the words of Miguel Gongora Meza: '*The coca plant is not cocaine...coca is an ancient medicinal and spiritual plant, initially domesticated about 5,000 years ago ...consumption of this plant as a food source is well documented...coca leaves supply high amounts of protein, vitamins and minerals...chewing the leaves does not cause any stimulant or euphoric effects nor does the practice cause any dependency*'.

The site of the city was occupied by the Aymara Indians and the Inca in pre-Columbian times; it was adopted by the Conquistadors in 1540. Many of the Spanish buildings in the city centre were constructed with white volcanic *sillar* 'rock' worked in *ashlar* style (ie 'finely cut masonry'), hence the name 'White City'. Arequipa claims 300 days of sunshine annually, with maximum temperatures of 25°C; the wet summer season is December to March, the remainder of the year being predominantly dry. The telephone code is (0)54.

Accommodation Etc. A full range of hotels is available, appropriately for a city of this size. Libertador (tel: 215110 – slightly off-centre but with extensive facilities to restore the balance), Sonesta Posadas del Inca and Casa Andina head the rankings (at a price). Mid-range, Casa de Mi Abuela (tel: 241206 – Jerusalén 606) fits the bill very well (swimming pool and extensive well-tended gardens), as does La Casa de Margott (tel: 229517 – Jerusalén 304), albeit more modestly.

When dining, keep in mind these Arequipeñan culinary specialities: *trucha*, and *chupe de camarones* – trout from the river Chili, and 'milky' soup of shrimps (likewise from the river); *cuy chactao* – roast guinea pig; *rocoto relleno* – local giant chilli (hot pepper look-alike) stuffed with meat, cheese and olives; *ocopa* – peanut-flavoured sauce to go with potatoes; *soltero de queso* – cheese-based vegetarian salad; *adobo* – pork chop soup; *pastel de papa* – potato cake (sliced layers of potato baked with eggs and cheese)

What to do and see. For tourists, the city is the launch pad for the **Colca Canyon** and the condors (see below), and possibly for **Cusco/Machu Picchu** and **Puno/Lake Titicaca**, and for trekking, cycling, white water rafting, and climbing **Misti** and **Chachani** (6057m/19,872ft).

But before all that, give consideration to the following: the **Plaza de Armas** (obviously) and particularly the magnificent cathedral (and museum) dating from 1612 (and successively restored after each and every subsequent earthquake, built predictably with the white *sillar* 'rock'); **Santa Catalina Convent** (Santa Catalina 301) - in a word, stunning; **Museo Santuarios de Altura** (La Merced 110)(including celebrated *Juanita* the preserved mummy); ***Mercado* San Camilo** (boxed by Piérola, Perú, San Camilo and Alto de la Luna); **Museo Histórico Municipal** (Plaza San Francisco); **La Recoleta Franciscan Monastery** (Jr.Recoleta 117), an under-visited historical treasure trove set in beautiful cloisters with a priceless library.

Also, cross Puente Grau to reach nearby (2km) **Yanahuara** for the church and mirador, and to visit **Museo Pre-Inca de Chiribaya** to see the important textile collection; a taxi will deliver you to **La Mansíon del Fundador**, the restored house and museum of the founder of Arequipa. For handicraft shopping, the **Plaza Handicraft Centre** (*Galeria de Artesanias "El Tumi de Oro"* – Portal de Flores 126) is the spot. (And for your restorative coffee and cakes, keep in mind nearby **Café Capriccio** at Mercaderes 121).

Moving on. The freight railway running between the port of Mollendo and Arequipa was opened as long ago as Christmas Eve 1870, the work of celebrated engineer Henry Meiggs. However, the War with Chile, the death of Meiggs in 1877 and financial problems conspired to delay final completion and inauguration of the Arequipa to Juliaca, Puno and Cusco lines until 1908. However, by 1893 the line was in service as far as Sicuani, although reaching Checcacupe (the next and penultimate station) took until 1906.

In the 1970's it was still possible to buy a ticket to Buenos Aires from the Station booking office in Arequipa, crossing Lake Titicaca from Puno to the Bolivian port of Guaqui in steamship *Inca.* From there trains run on the 1903 track to La Paz, and onwards to Argentina, thus completing the link from the Pacific to the Atlantic. Until very recently there was no current passenger service on the Arequipa to Puno portion of the railway, which had become exclusively used for freighting minerals (copper ore, cadmium), nor directly across the Lake to Guaqui. However, thanks to Belmond, owners of the Andean Explorer, Orient Express and other lines, that has now changed – and, who knows, possibly SS *Yavari* (dating from 1862) may soon be licensed to carry passengers once again from Puno to Guaqui.

Buses run regularly from **Arequipa** to **Juliaca/Puno** (4-6hrs) from Terminals Terrestre and Terrapuerto by both day and night, and onwards to **Cusco** (10-11hrs). Bus journey time from **Arequipa** to **Chivay** (171km) can be as little as 3hrs, extending to 4 or even 5hrs in bad weather (snow, fog, rain). Amongst the competing operators, *Empresa* Reyna depart daily from Terminal Terrestre at 1100, no frills, S/.13.00; baggage reclaim can be chaotic, loading and unloading takes place simultaneously on your arrival at the bus station in Chivay (3 blocks from the Plaza), with predictable results – keep a sharp eye to avoid losing your luggage.

As a footnote, it is possible to undertake a one-day excursion from Arequipa to Chivay and the *Cruz del Cóndor Mirador*, setting out at 0400 and returning at 2100 – a longish day on the road, especially at an altitude of 3,600m/11,811ft, which may not suit everyone. Alpaca, llama and vicuña may be seen en route, together with customary Andean scenery.

Courtyard, Recoleta Monastery

Peru - as perceived by the Missionaries of the Colonial Period

Recoleta bell tower

The priceless collection in the Recoleta library

Intricate Andean Handiwork

Chivay

At a height of 3,600m (11,811ft) Chivay is the largest of several delightful Colca towns/villages, including Yanque, Maca, Coporaque and Achoma, and is the recognised access point to the Canyon. There is a bridge here over the río Colca, linking both sides of the canyon.

Plaza de Armas, Chivay

There are numerous accommodation options in Chivay. At Casa Andina (tel. 531020) (5 mins walk from the Plaza) is the Maria Reiche Observatory with an astronomy telescope and presentations for visitors (closed December to March when the sky is cloudy). The Colca Lodge across the river from the nearby village of Yanque takes advantage of the hot springs in the grounds.

Andean figures, *Calle* Alameda

Leading from the Plaza, *Calle* Alameda with its colourful statues of Andean figures and costumes runs into Av. Salaverry and the Chivay market area, worthy of a visit for handicrafts. Cycles may be hired from Sr.Wilde, *Bici Sport*, Zarumilla 712.

Inca Bridge Chivay

Continue to the end of the street to see where the Colca river is spanned by the iconic Inca bridge dating from pre-Hispanic times. In the main square of Chivay are several restaurants, including *El Colonial*, and at 301, *Aromas Caffee*. *Pique* (aniseed and hot wine) is an Andean speciality.

Street scene

Private (S/.160 for two) or bi-lingual group tours (S/.50 pp) to the Cruz del Cóndor Mirador may be arranged with Tourist Agencies on the Chivay plaza, for example with Peru Travellers (Javier & Ruth – Plaza 105), departing from 0600 onwards daily, with hotel pick up, and returning before lunch (plus other activities: camping, trekking etc).

The Colca Canyon

The Colca Canyon is home to the Andean Condor ("Messenger of the Gods"), a Peruvian national symbol closely entwined in the folklore and mythology of the mountain regions. The viewing point for these majestic but now endangered birds is the *Cruz del Cóndor Mirador*, about one to one and a half hours by road from Chivay and three quarters of an hour from Cabanaconde at the far end of the canyon. There is ample space for the many visitors; a tourist toll charge of S/.35 per head is levied.

Canyon topography

Opinions vary on the best time to see the Condors, but the consensus is 0800-1000 daily. The dry season of May to December is optimum with perhaps 30 birds (the colony has reputedly risen to around 50), although January to April is perfectly feasible, albeit with a much reduced number of sightings (and very occasionally the road is impassable). The depth of the Colca Canyon is 3270m (10,725ft), twice that of the Grand Canyon, and just marginally less (163m) than nearby Cotahuasi Canyon (the deepest in the world). A feature of the Canyon are the extensive terraced fields created by the Wari (xxxxx) long before the arrival of the Conquistadors and still cultivated today by the Collagua and Cabana people. *Las Tumbas Colgantes,* the hanging tombs found throughout the Canyon (and near to Chivay at Yuraq Qaga and Choquetico) have the same ancestry. During the rainy season the Canyon is particularly colourful.

Condor feeding

A Condor's wingspan can reach 3.3m (almost 11ft), exceeded only by two sea birds, the Albatross and the Pelican. They lay just one or two eggs per year (reputedly the size of a human head), but have a possible lifespan between 30 and 70 years. Condors are monogamous; when one of a pair dies, the surviving mate will commit suicide by diving into the ground from a great height. Condors have a symbiotic relationship with foxes, notifying finds of food.

Colca vendor

In Praise of the Condor

Hoy, el 'Condor Pasa', pasó por mis ojos.
Al verte mi corazon latió, latió ciento de veces
Pidiendo que una vez mas te dejes ver con tu majestuosa belleza
Que acarician los vientos mas profundos de nuestros sagrados Apus.

Condor, porque eres el mas esplendoroso ave de todas las aves?
Ven, ven y acaricia con tus plumajes mis sueños andinos
Como una tierna nube del amanecer que va por encima de las montañas
Besando y acariciando con los Dioses del mas alla.

Esta noche me embriague en tu belleza al pensar que estas ahí y siempre estaras ahí!
Viva la hermosura de tu ser y que al volar lleves contigo los sueños de tus hijos
Los viajeros que vienen y que se van
Majestuoso Condor, vuela y vive para siempre en nuestros corazones.

****

Por la primera vez hoy, mire tu majestuoso vuelo entre los Apus mas bellos de nuestro Perú
Eres lo mas asombrosa ave entre todas las aves y aun asi tus delicados plumajes
Acarician los vientos mas profundos de los Andes
Mi bello Condor, al pasar tu infinito vuelo sobre mi lejana mirada te mire fijamente a
Tus ojos y te pido que una vez mas me lleves entre tus alas para volar juntos al mas alla sin retornar.

****

Condor, tus plumajes blancos son como los manos de un Angel que fragilmente puede tocar los Vientos del mas alla en una mañana de cielo azul que va desasiendo el rocio del amanacer
Condor mi eternal Condor, amigo de mis suenos, aquel quien puede llevar en sus alas
Las alegrias y las tristezas de los pueblos que te vieron nacer
Inspirafortaleza, poder, fragilidad, libertad y belleza inalcanzable por el inferior que te observa.

Eres la pura sangre de tu pueblo, el que te acompanan desde la fundacion de sus hijos
Solo podemos respirar el aire que desechan tus delicados plumajes
Condor, eres aquel quien nadie puede tocar, vive y vuela para siempre
Para siempre seras majestouso para cuidar de tus generaciones y para embellecer los cielos y las tierras santas
Condor, Condorcito, eres el alma del presente y el pasado para el Regocijo de aquel que te pueda ver y amarte para todo la vida.

****

Finalmente hoy en un mes frio de enero me llevo en el alma lo que era buscando en este lejano pueblo
Con tu viva mirada tocaste mi alma y corazon que latia intensamente al ver pasar tu elegancia en mi caminar
Condor de los milenios, como el aire que viene y se va te llevo en mi para siempre
Como un sueño de nunca despertar

****

Plaza Pino, Puno

Puno Cathedral 1857

Central Market

Traditional costumes

Arco Deustia

Festival costumes - Av.Laykakota

Puno

Getting to Puno. The bus companies in the Chivay road terminal are principally focussed on Arequipa. However, three tourist services (with fares to match) run to Puno from Chivay, a distance of 274km covered in 6hrs – more in conditions of fog and rain/snow. Operator 4M Transporte Turistico (HQ in Arequipa) is based on La Pascana Hostal, just off the corner of the Chivay Plaza; their bus (which comes in from Puno) leaves at 1300 daily, price USD $50.00 per passenger. *Rutas de Sur* and My Tour Peru prices are $46 and $35 respectively. Bookings may be made with Peru Travellers on the Plaza (105).

Travelling with 4M there is a meal break and stops at viewing points en route, the first being at Patapampa not far from Chivay, which at a height of 4,800m/15,748ft provides splendid vistas of the volcanoes and snow-covered mountains on a clear day. You will also see dozens of ritual cairns (*apachetas* in Quechua) – individual piles of rocks assembled by travellers to assure a safe journey in good health. Further on, the road passes through Pampa Cañahuas, part of the Aguada National Park where the endangered vicuña may be sighted. On reaching the Department of Puno a stop is made at Lagunillas Lake at 4,174m (13,694ft) to see the flamingos, *ajoyas* (scavenging ducks) and the *choca* Andean duck. A bi-lingual guide provides an informatory commentary throughout the trip – arrival time variable between 1900/2100.

Buses run from Arequipa to Puno via Juliaca (5hrs); buses from Lima stage through Arequipa (21hrs and upwards). Bus journey time from/to Cusco via Juliaca is 5-7hrs; tourist buses with scenic stops and lunch can take 10hrs on this road. Regular air services run from Lima to Manco Cápac airport in Juliaca, 1hr by road from Puno; flight time is 1hr 40mins or via Arequipa 2-3hrs. Avianca run two flights daily; LATAM (Lan Chile + Tam Brazil = LATAM as of May 2016) run four flights per day (2 via Arequipa).

Potted Profile. With an estimated population of some 150,000, Puno lies on the western shore of Lake Titicaca (dealt with separately below in following sub-section) and is (logically) the capital of the department of that name. It is the acknowledged *centro folclórico* of Peru and the costumes, dance and music in the festival parades throughout the year (especially before Easter, and also the *Fiesta de la Virgen de la Candelaria* in February – with rehearsals starting in January) are spectacular. At an altitude of 3,855m/12,647ft it is the 5th highest city in the world; the night temperatures are regularly well below freezing, particularly so during the southern hemisphere winter months of June to August. The telephone code is (0)51.

Accommodation Etc. Well appointed and well recommended Hotel Hacienda lies directly on the Plaza (Puno Plaza 425) (tel: 419425) – not to be confused with Hacienda Puno at Jr.Deústia 297. Also on the Plaza is the smaller but excellent Hotel Colonial (Jr.Puno 529)(tel: 363928), and just up the slope at Jr.Puno 681 is the modern Conde de Lemos (tel: 369898). Hotels Libertador and Sonesta, and Casa Andina are on the Lake waterfront just to the north of Puno, 4-5km from the city centre.

Jr.Lima running from the Plaza de Armas to Plaza Pino contains many restaurants; for example, at No. 394 is Tulipan's and at No.517 is the Hacienda Restaurant/Pizzería. A popular speciality of the Puno region is *Api*, a hot drink made from purple and white *maíz* (sweet corn) flavoured with cinnamon and sugar, served with cheese empanadas or bread in dedicated *Api* cafes – a delicious 'teatime' confection taken at any hour . Also, Puno *chicha* is often made using *quinoa*, rather than the customary purple *maíz*, and be sure to try hot maracuyu juice and *Crema de Chuño* soup based on potatoes that have been frozen at altitude and stored.

What to do and see. As is customary, on the main Plaza is the grand Colonial cathedral, dating from 1657. Just off the Plaza at Conde de Lemos 289 is the excellently presented **Municipal Museum**, with a wide variety of gold, silver and pottery artefacts of interest, open daily from 0900-1900 (S/.15.00). The accumulated historical possessions (pre-Hispanic, Colonial and ethnic) of German-born painter and collector Herr Carlos Dreyer (1895-1975) who lived in Peru for 50 years are now in the museum, plus many of his paintings, hence *Museo Municipal Dreyer.*

There is a small **naval museum** (free) at the corner of Av.Titicaca/Av.El Sol (0800-1700 Mon to Fri; 0900-1300 Sat/Sun) dealing with navigation and activities on Lake Titicaca, but which could be expanded to advantage cover more history of naval operations thereon in past years. At the end of Av.Titicaca where the boats leave for the Islands there is an artificial boating pool formed by the construction of the outer mole, with **pedalos** for hire, which may or may not appeal.

Puno's **Central Market** is just two blocks East from picturesque Plaza Pino at Oquendo 154 and bounded by Tacna, Arbulú and Arequipa: colourful and varied, as are all central markets in Peru, with the usual wide selection of food stalls. Handicrafts stalls line the road down to the lakeside port, while ***Mercado* Laykakota** (South along Jr.Arequipa to Av.Laykakota) is unbeatable for local produce, both grown and crafted.

Whilst in the vicinity of Laykakota, visit fascinating sculptural ***Cementerio Central*** and the shops specialising in the production of the lavishly decorated flamboyant Festival costumes. Then proceed to **Arco Deústia** up Jr.Independencia to see the monument constructed in 1847 in memory of the fallen at the definitive independence battles of Junín and Ayacucho in August and December 1824, before visiting the immediately adjacent *mirador* to overlook the town and lake.

There are several excursions of architectural and religious significance to be undertaken in the vicinity of Puno, not least being a half-day tour to the *chullpas* at **Sillustani** (32km). These pre-Colombian funeral/burial towers dating from the 15th century are ascribed to the Colla tribe, and the complexities of their construction have confounded latter-day archaeologists.

Moving on. As mentioned earlier, the Puno to Cusco railway line, via Juliaca, opened in 1908. The train station and Peru Rail ticket booking office (open 0700-1700 weekdays, 0700-1200 Sat/Sun) are on Av.La Torre; you can also book online at perurail.com and there is a call centre on 51.84.581414. The Andean Explorer service leaves at 0800 on Mondays, Wednesdays and Saturdays, arriving in Cusco at 1830. Quoted price $316 – far from cheap, but a rare experience. Book in advance.

La Paz is the next planned destination for many visitors to Puno. Crillon Tours and Transturin (agent at Jr.Cajarmarca 287 on cel.956054500) use hydrofoils and catamarans respectively to cross the Lake. They offer a variety of trips by road and water through Copacabana (Bolivia), plus sightseeing en route – both agencies are on line for full details and bookings. The relevant border point for immigration in this case is at Yunguyo.

Bolivia Hop ('Bolivia's first hop on hop off bus system') have Puno to La Paz buses for $49 (suggested minimum time required: 14hrs), plus other combinations advertised online (no maximum times for these variable excursions). Bus companies Panamericano (leaving Puno 0730) and Tour Peru (also with daily departures) both run through Yunguyo (border control) and Copacabana to La Paz, taking around 7hrs (seats on the left hand side recommended for scenic views of the Lake). Economic buses for the direct route to La Paz (256kms/5hrs) are available from the Puno Terminal Terrestre (eg Ormeño departing 0545 daily, and others), passing through the Peruvian border office at Desaguadero.

You can fly daily from Juliaca to Cusco in under an hour, but the road journey from Puno by bus (388kms) can be achieved in 5-7hrs and is more rewarding for the traveller. Lots of daylight options from Terminal Terrestre eg Libertad and Power for S/.25.00, up to a 10hr tourist excursion with Inka Express for $60.00. Points of interest on the way (fully paved and generally running in company with the rail track across the *altiplano*) are pottery bulls at Pucará, the pass at La Raya (4321m/14,176ft), the town of Sicuani, followed by the scenic valley of the river Urcos, and the village of Andahuaylillas (possessor of a 17th century church sometimes termed the 'Andean Sistine Chapel).

Of particular note on the Puno to Cusco road is that some 35kms past Sicuani, over to the right looking towards Pitumarca, can be seen the snow covered peak of Mt Ausangate. At 6,384m/20,945ft this is the fifth highest mountain in Peru, situated in the Vilcanota range. It is especially significant in that the area is the remote home of llama and alpaca herders, a rare surviving untouched pastoral society; moreover, the geology of the region produces singular and remarkable rock colourings, and nearby the annual festival of the 'star snow' is attended by thousands. Treks to the region are available through *Apus Peru.*

Fiesta de la Virgen de la Candelaria

Lake Titicaca & *Los Uros*

The largest lake in South America. At a height of 3,812m/12,507ft, Titicaca has assumed the title of 'highest navigable lake in the world'. The eastern side is in Bolivia, the western portion is Peruvian. At its widest point it is 50 miles across; it is 118 miles long, with an average depth of 351ft. It covers 3,232sq.miles (ranked 18th in the list of largest lakes in the world).

Titicaca - viewed across the Puno rooftops

As is the case with so many other features of Planet Earth, Lake Titicaca (and the unique wild life therein) is under environmental threat. The water level is falling, occasioned by reduced annual rains and the receding glaciers that supply the tributaries that feed the lake. Moreover, water pollution is a serious threat to wildlife and lake dwellers alike, caused by the expanding towns and cities adjacent to the shore, coupled with gross debris contamination.

Legend has it that **Manco Cápac** and **Mama Ocllo**, founders of *El Imperío Inca* in times long past, originally arose from the waters of Lake Titicaca. However, earnest latter-day scholars in the field of mythology are wont to cast doubt on this occurrence.

'pejerrey'

Uru family

The floating homes of **Los Uros** (Uru people – Amazon migrants) made from the buoyant *totora* reeds of the Lake are nowadays a major tourist attraction. Each artificial island (60 in 2000, now numbering over 100) nominally contains one extended family housed in several dwellings, living on fish (*trucha, pejerrey, carachi, mauri, camarones*) and income from handicrafts. Boats to visit the Uros run from the end of Av.Titicaca in the Puno port (S/.10.00). You may also stay overnight with the families; solar panels provide energy, but take drinking water; fruit is always welcome (S/.30.00 pp negotiable – double through your hotel/an agent).

Onshore for shopping

Reed dwellings

There are several inhabited islands on Lake Titicaca. **Taquile** (pop.4,000) and **Amantani**, both Quechua speaking and 45km from Puno port, are packed with ruins of interest, plus world renowned woven textiles and woollen handicrafts (the men knit, the women spin and weave), and a traditional clothing museum on Taquile. Aside from fiestas, the islands are very peaceful; you can stay overnight (not least to enjoy the stars). Way to the east on the Lake (*Lago Menor*) is the little Peruvian island of **Anapia**, reached by boat from Yunguyo; the Aymara speaking residents welcome visitors to stay. (*Islas del Sol* and *de la Luna* are in Bolivian waters; likewise *Suriki*, noted for reed boat construction).

'Toothpaste' reeds

Handicrafts

The Titicaca Steamships

The first substantial vessels to operate on the Lake were the ***Yavari*** and ***Yapura***, gunboats for the Peruvian Navy with cargo/passenger capacity. London-built in 1862 by Thames Ironworks in kit form, the 2,766 boxes were carried by men and llamas from the port of Callao, Lima up and over the Andes to Puno. Yavari was finally launched on Christmas Day 1870, Yapura in 1872.

Yavari underway

The ships, named after rivers in Loreto, were 180 tons, 30 metres in length and steam driven, using llama dung instead of coal. Diesel engines were fitted in lieu in 1914 and the hulls lengthened. In 1975 *Yapura* was renamed BAP Puno by the Peruvian Navy, and is still operating on the Lake as a hospital ship. *Yavari* was saved from the breaker's yard in 1987 via a conservation project (directed by Ms Meriel Larkin, descendant of the founder of Yarrows Shipbuilders), and is now a museum ship, offering overnight accommodation anchored off Hotel Libertador. The aim is to enter commercial passenger service as the world's oldest – and highest – working single-screw iron ship.

Yavari lying at anchor 2017

Yapura - Coya - Yavari

(Foto: Delgado de la Flor)

The next ship to arrive on the Lake was the ***Coya*** ('Princess' in Quechua), built 1892 by Wm. Denny & Bros, Dumbarton, River Clyde, Scotland and launched 1893 with a length of 52m and 546 tons – supplied in pieces, well timed for freighting to Puno on the brand new railway from the coast at Mollendo. Beached in 1984, she serves as a restaurant whilst being restored by a Peruvian consortium.

SS Coya beached - 2017

SS Inca, 67m and 1,809 tons arrived by rail in kit form in 1905 from Earle's, Kingston upon Hull, Humber. She gave service for 90 years between Puno and Guaqui, being broken up in 1994 – well remembered by many past travellers.

SS Inca - the way she was

(Foto: Delgado de la Flor)

At 79m and 2,200 tons ***SS Ollanta*** (renowned Inca General) was the largest and final of the five Lake vessels, the kit pieces coming from Earle's via Arequipa in 1930. No longer in service, property of Peru Rail, she is berthed in Puno port, hopefully destined for restoration.

SS Ollanta - in her prime

Ollanta 2017 - awaiting restoration

Ollanta - passengers and freight to and from Bolivia...

Representation of Cantuta
'Sacred flower of the Incas
and National Flower of Peru'
found on the altiplano *yungas*

Apachetas - the 'safe' journey
cairns of the Andean travellers

Implements for Andean back-strap weaving

Andean textile design

Artistic portrayal of the
Andean sun god

Pan pipes of the
mountains (*siku*)

Section II
The Central Highlands

"From the mountains of Peru you may kiss the sky, and put out your hand to touch the face of God"

The Railway to the Moon

This poetic description of the epic Callao to Huancayo Central Railway of Peru (also styled 'The Railway in the Clouds') has been much borrowed, but it was, and is, a truly staggering feat of engineering over a challenging mountainous distance of 332/345kms, finally comprising 68 tunnels, six zigzags, and 61 bridges (not least being the Verrugas Viaduct at Km 84, 755 feet in length, 252 feet above the gorge; built 1870, swept away by flood waters 1889, re-built 1891, swept away again in 1937, repaired and re-named the Carríon Bridge in 1938, blown up by Sendero Luminoso in 1993, repaired).

First conceived in 1859 as a track from Callao to La Oroya for the carriage of minerals and ore, the first route surveys were conducted by Polish engineer Ernesto Malinowski. American Henry Meiggs, already at that time engaged on construction of the Mollendo to Arequipa line (and credited with the precept of 'I will located the rails wherever the llamas go', sometimes rendered as 'I will place rails there, where the llamas walk') then offered an alternative complementary survey, albeit based on the work of Malinowski. The upshot was that construction under the aegis of Meiggs started on 1st January 1870. Unfortunately, Meiggs died in 1877, and with that went his dream to 'scale the summits of the Andes and to unite with bonds of iron the people of the Pacific and the Atlantic'.

Floods, financial problems and the War of the Pacific also conspired to delay progress, but by November 1893 a distance of 221kms had been completed and the way to La Oroya was open. The Galera Tunnel at 172kms is the highest altitude attained by the Central Railway, 4,770 metres above sea level (15,681ft claimed – alternative records quote the high point as being La Cima at 4,835m/15,863ft), making it for more than a century the highest railway in the world, only recently exceeded in 2006 by the Qingzang/Qinghai-Tibet Railway which crosses the Tanggula Pass at 5,072m/16,640ft. In 1904 the extension of

the line from Oroya to Huancayo was approved, and on 24th September 1908 the end of the line was reached at last after 38 years, with 206 miles of Standard Gauge track (4′8½″) in place.

Steam locomotives from USA were first used to haul the assorted cargos of minerals, fuel, food and cement on the line, but these were soon replaced by specially designed oil-fired 'Andes' engines from the firm of Beyer, Peacock and Company of Manchester, UK (2-8-0s for the purists). These in turn were overtaken by ALCo (legendary US train maker, closed 1969) diesels in 1963, lasting until 2006 to be superseded by General Electric locomotives fuelled by compressed diesel and natural gas.

Initially run by The Peruvian Corporation, operation of the Ferrovías Central Railway (Central Railway of Peru) was assumed in 1972 by Peruvian National Railways (ENAFER – Empresa Nacional de Ferocarriles del Peru, who were in turn replaced by FCCA (Ferrocarril Central Andino) in July 1999.

Twenty years ago a weekly passenger service from Lima to Huancayo was still available for S/.30.00 for what was then a 12hr journey, bottled oxygen being on hand for use as necessary at the higher altitudes. The current situation is that FFCA run twelve tourist services annually, starting after the rainy season with two trips in April, one in May, two in June etc through to end of season November. Dates are notified online at www.ferrocarrilcentral.com.pe where bookings may be made; fares (turístico/clasíco) are currently in the range of $212 to $303 for what is now a 14hr excursion with two scenic stops and full facilities (including the oxygen plus nurse), the train leaving the Desamparados Station in Lima at 0700 and reaching Huancayo at 2100.

(Trevor H.Stephenson)

Desamparados Railway Station, Lima

(Trevor H.Stephenson)

Verrugas Viaduct, 1891

Huancayo

Getting to Huancayo. As an alternative to the train, buses from Lima take about 6-7 hours to cover the 335kms, the paved road following much the same route as the Central Railway of Peru (as above). However, the advertised travel time can extend to 9 hours in conditions of rain, mist and snow (most likely to be encountered from December through to March), the route reaching almost 4,800 metres (15,748ft) over the Ticlio Pass just beyond the halfway mark. As is invariably the case in the Andes, the scenery is spectacular and travel by day is recommended. Initially bordering the river Rimac, from the attractive little resort town of Chosica (40km from Lima, and clear of the garua mist) the road climbs via a succession of green hills and valleys to progressively barren mountain terrain, finally passing the desolate mining centre of La Oroya (3,755m/12,320ft), two thirds along the way). From here the road runs on to Huancayo, descending only marginally to 3,271 metres (10,732ft) to arrive in the city.

One of the many Lima departure options is from the San Borga terminal of Movilbus leaving at 1230 daily (S/.65.00 single, panoramico seat). To fortify you for the journey, try an Andean *ponche* on sale at the Terminal while you wait, as imbibed by the Incas – perhaps one with abas beans and ground maca tuber, or alternatively a quinua grain cereal.

Potted Profile. At an altitude of 3,271 metres (10,732ft), it was here that the presidential decree abolishing slavery in Peru was signed in 1854. The population of Huancayo has more than doubled to over half a million since the turn of the century, and the growth in size of the city has been such that construction of a commuter railway was proposed in 2012 (the Huancayo Metro – yet to be implemented). As the principal commercial hub for the Central Highlands, rapid expansion and urbanisation has flowed somewhat indiscriminately into the surrounding scenically fertile Mantaro Valley, memorable amongst other things for the profusion of the yellow retama shrub in the hedgerows (a flower poetically linked with the eternal struggle for independence). However, the rich culture of the area is still to be encountered in the city's very extensive daily and weekly markets of local produce and handicrafts, and in nearby villages, and in the customarily very full calendar of colourful Festivals. Telephone code: 064.

[Intensive agriculture, discredited European vogue of the 1950's, complete with massive applications of fertilisers, has been introduced widely throughout the Andes by overseas 'experts', superseding the centuries-old wisdom of the

Incas and of earlier civilisations, and replacing rotational cropping. Not only has this been to the detriment of the environment but there is strong anecdotal evidence to the effect that the incidence of cancer cases has markedly increased and life expectancy has fallen.]

Accommodation Etc. The Kiya Hotel on the Plaza at Av.Giráldez is a good mid-range choice (tel: 214955). More expensively, Hotel Turismo on Av.Ancash 729 is also conveniently central and has a good restaurant (tel: 231072). And less expensively, Casa Alojamiento de Aldo y Soledad Bonilla is a colonial house with courtyard at Av.Huánuco 332 (tel: 232103) just six blocks from the Plaza.

What to do and see. The city Museum at Pasaje Santa Rosa just across the river Shulicas to the north is an eclectic mix of curiosities, worth combining with a visit to Parque de Identidad Wanka, north east Huancayo for historical and cultural information about the Mantaro Valley and the trees and plants. Away from the city, the Valley itself is full of interest with many nearby village excursions to be undertaken within a radius of 10 to 15kms. Such as to San Jerónimo for silver filigree jewellery; Hualhuas for woven alpaca; San Agustin de Cajas for hats; San Pedro for wooden chairs; Cochas Chico and Grande for carved gourds (and fine views to be enjoyed).

Further afield, at 60kms (but not much more than an hour and a half by road), is Izcuchaca (halfway along the railway to Huancavelica – see below) where there is a Spanish bridge over the river Mantaro and a water mill pottery. Reverting to Huancayo itself, the central market (Mercado Modelo) is (as always in Peru) excellent for food stalls, whilst the Ancash/Real handicraft market is extensive, as is the even bigger market on Jirón Huancavelica which sells everything imaginable.

Moving on. The Terminal Terrestre for most buses (some companies have their own – Cruz del Sur invariably, for example) is just north of the city centre. There you will find services to the south east to Huancavelica (5hrs) and Ayacucho (10hrs). Going to the north west there are buses to Huánuco (7hrs); Jauja is just 44kms along this road, and beyond there lies Tarma (see below for both these destinations). The railway may of course be taken back to Lima from the main city station, if the FCCA monthly schedule fits your programme; however there is a regular train service running to Huancavelica from the entirely separate station in the Chilca suburb of Huancayo, and this is highly recommended – see below for full details.

Jauja

Situated in the Mantaro Valley, easily reached by *colectivo/combi,* 1hr from Huancayo, S/.3.00. Or fly from Lima with LC Peru (2 flights per day). Population now 30,000, altitude 3,400m (11,155ft).

One of the first Conquistador cities, Francisco Pizarro's choice of capital before Lima in 1535. There are Inca storehouses to be seen above the town, and scenic Paca Lake is just 3.5km away. The Gothic-style Cristo Pobre church overlooks the Plaza and is said to owe its design to Notre Dame.

Cristo Pobre church

The healthy climate of Jauja prompted foundation of a cosmopolitan sanatorium after the end of the War of the Pacific in 1883, dealing particularly with cases of tuberculosis. In the War itself, Mariscal Andrés Cáceres ('Wizard of the Andes') was active in this region.

The calendar of assorted festivals celebrated in Jauja runs right through the year.

Tarma

Founded 1534, formerly 'tranquil Tarma, Pearl of the Andes and City of Flowers'; expanding now, population 60,000; altitude 3050m/10,007ft. An hour from Jauja in local transport, or bus from Lima over the Abra Anticona -Ticlio Pass (4,818m/15,807ft) Worth the journey to see the beautiful terraced countryside and 'flower fields filled with industrious campesinos tending gladioli, gypsophila, carnations, stocks, statice – and spinach'. Here, locally made flower-carpets are a speciality.

Stay in Hotel Los Portales (064.321411 – Av. Castilla 512) for a blazing fire of eucalyptus logs ('eccentric hospitality, bizarre plumbing'). Hacienda Santa Maria (Vista Alegre 1249, Sacsamarca) (064.321232) is 17th century with extensive gardens.

To complete your transit of the Andes, take the road eastwards from Tarma to La Merced (alt.751m/pop.25,000). In 80km descend 2,450m to pass through 'the sweetly scented orange and tangerine plantations of the tropical Chanchamayo Valley, home to the Asháninka' (bows, arrows, necklaces on sale). Oxapampa (pop.70,000) and Pozuzo (10,000) follow, founded by Austro-German settlers 150 years ago, as evidenced in the architecture and local customs of the present-day farming communities.

El Tren Macho

Introduction. The railway line from Huancayo to Huancavelica runs for 128kms - or originally for 148kms, including the now defunct extension to Lachoc where the local senator lived at the time of inception (sic). Run by ENAFER (Peruvian National Railways) from 1973 until its disbandment in 1999, it remains today one of only two railways in the country still belonging directly to the government of Peru. Linking remote Huancavelica with the pulsing commercial hub of Huancayo, and beyond that with Lima, the line is a hugely beneficial economic artery for the region.

The service is affectionately known as ***El Tren Macho*** because at one time it was plagued by recurring technical problems, and it was said that the train 'departed when it wanted and arrived when it could'. History now, and today it operates with clockwork punctuality.

History. The original 1904 concept was to connect the Peruvian Central Railway with the Southern Railway (both mentioned earlier herein), albeit with a 'narrow gauge' (3 feet wide, as opposed to 4' 8½") track, following a circuitous route to Ayacucho. Work started in 1907 but stopped in 1910 with only eight km of track laid, and 20 km of roadbed prepared. It was not until 1918 that construction was restarted, although with a modified plan: from La Mejorada at km 78 a branch line was to go to Huancavelica, and the main track would continue to Ayacucho. In the event the latter portion of the project never came to fruition – if it had, the 4,960m (16,273ft) section at Castrovirreyna would have become the highest railway line in the world at that time, overtaking the Central line altitude of 4,810msnm.

Various track enabling preparatory works and tunnel entrances may to this day be seen from the road between Huancavelica and Ayacucho but frequent landslides and other obstacles finally dictated cancellation of the endeavour. As it was, the section of the line up to station Maríscal Cáceres (km 76) opened for traffic in 1926, and the complete route to Huancayo opened in 1933, exactly as is operational today. With just one modification: the original 3ft 'narrow gauge' track was converted to 'standard gauge' (4' 8½") in 2010 to enhance capacity and efficiency.

Specifications. The 128km line has 15 bridges and 38 tunnels, traversing the Rio Mantaro Canyon to the junction of the Rio Huancavelica. Highest point 4,700msnm or 15,420ft. Stations along the way are: Chilca (Huancayo); Manuel Tellería; Izcuchaca; Maríscal Cáceres; Acoria; Yauli; Huancavelica. The original US Baldwin (based on Leeds, UK design) steam locomotives have long since been replaced, although one (No.107 made in 1936 by the Leeds Hunslet Engine Co.) may still be seen preserved at Chilca station. The current engines are diesel and natural gas.

Today. This is a vintage Andean train journey, not overlong at seven hours (but susceptible to landslip and cancellation in the rainy season), on a legendary railway. The adventure may surely be ranked as one of the most picturesque (and even one of the great) rail journeys of the world, not only on account of the mountain scenery, but also for the colourful traditional attire and company of one's fellow travellers.

The four carriage *El Tren Macho* departs at 0630 sharp on Mondays, Wednesdays and Fridays from Chilca Station (just 10mins from Huancayo centre). Go to the station the previous day to buy your tickets: comfortable Primera is S/.9.50 and Buffet (reserved seating) is S/.13.00 – this is a working railway much utilised by the community; inflated tourist prices as found on other routes have yet to appear... The onboard kitchen and vendedores ensure an unlimited supply of food for your table, completing an unmissable experience. It could even be said that *El Tren Macho* is one of the great culinary experiences of Peru – be sure to sample the Pan de Calabaza pumpkin bread, and the Agua Muña herb potion. But if needs must, there is also a daily 1230 Auto-Wagen three-hour express service (at 1400 on Sundays) between Huancayo and Huancavelica.

Arrival in Huancavelica

Huancavelica

Getting to Huancavelica. As mentioned, easy and scenic 5hrs/147kms on paved road by daily buses from Huancayo, but *El Tren Macho* is the best option. There is also a nightly bus link (12hrs) from/to Pisco and Ica on the coast.

Potted Profile. Altitude 3676m/12,060ft, population 50,000, phone code 064; dry but cold February to August, the rainy season starts in September. Often stigmatised with the description of being 'one of the poorest cities in Peru' – certainly not the case today, and a delight to visit, not least for the prevalence of traditional Andean attire and the beautiful surrounding mountains, high above which may be seen the occasional condor.

The city was formally established by the Spanish in 1572, although of course the area had been inhabited since long before that time, not least by the Wankas in pre-Incan days. The famed Santa Barbara mercury (quicksilver) mine was but 3km from Huancavelica, important catalyst for extracting the valuable silver deposits in Colonial times, not only in the region but also in distant Potosi in Bolivia.

Accommodation Etc. The 3-star Hotel Presidente (tel: 452760) commanding the Plaza de Armas cannot be faulted for position and warmth (open wood fire if requested). Hotel Victoria (tel: 555123/555119), corner of the Plaza (Manco Cápac/Virrey Toledo) is also 3-star: full facilities and a very reasonable tariff. Room heating is available on request in the hotels, but in budget hotel Ascención (also on the Plaza, and much colder since they egregiously replaced the warm wooden floors with slippery stone) beware the unannounced S/.20.00 daily charge that appears on the bill at check out.

What to do and see. The cathedral on the picturesque Plaza de Armas is renowned for its altar, and there are eight more churches nearby. Both on the Plaza, and in the street market just behind, local produce and handicrafts are on sale. Crossing *el río* Huancavelica and climbing the hill to the north takes you to the local thermal baths, plus providing a good view. There are other longer walks to be enjoyed (e.g. Potaqchiz hill, about one hour away).

The customary calendar of Andean festivals and pageants is to be encountered in Huancavelica, centred on the National Cultural Institute just 100m from the Plaza – not least and particularly colourfully over the Christmas period. There is a small regional museum (Av.Arica/Raimondi).

Huancavelica Plaza & Cathedral

Historic Santa Barbara Church

Street spinning, Huancavelica

Traditional costumes

Christmas Festivals in Huancavelica

National Cultural Institute

Steps to the Thermal Baths

River Huancavelica

Moving on. Back to Huancayo in the bus (or the train), then on to Lima 13hrs/445kms. Pisco/Ica – 12hrs/269kms. Molina bus company (Av.Celestino Manchego Muñoz 1004) run to Ayacucho at half past midnight nightly, S/.50/70 depending on fiestas etc. Does the job but you will forego the spectacular scenery. Think instead of finding a private car and driver; 247kms - ask around. For instance: ET Tours Nor Oriente, No.799 Av. CM Muñoz, tel: #945843669 – 3 pax/baggage S/.300 (Christmas Day - all things are possible).

Ayacucho

Getting to Ayacucho. The journey of 247kms from Huancavelica is on paved road, albeit quite narrow initially (the era of Sendero Luminoso mining the route against the military is long past). The steep 1000m climb twisting and turning to reach Pucupampa (km.43) at 4,500m produces spectacular views of snow-covered rocky peaks, and llama and alpaca herds (threatened by foxes and even puma). At km.55 the altitude reaches 4853m/15,922ft as the road traverses the Chonta Pass (Abra Chonta). Beyond Santa Inés (km78) wild duck and flamingo are seen on the altiplano lakes, as are many apacheta 'safe journey' cairns. Not to mention abandoned villages and dwellings (a legacy of the civil war). Travelling time 5-6hrs, depending on weather conditions, including falling and fallen snow, plus photo-stops – memorable moments are many.

Civa buses leave Lima from Av.28 Julio/Paseo la Republica daily at 1730 to arrive in the city of Ayacucho twelve and a half hours later at 0600 for S/.90.00. There are many other competing operators, e.g. Los Chankas, Molina, Ronco, Espinoza etc, also serving Andahuaylas (16hrs). The entire way is paved now; the epic era of travelling on the back of a truck through the dust and mud for several days is but a distant memory of a stoical childhood.

Early morning flights run seven days a week from Lima – Star Peru 0530/0630. LC Peru have a convenient schedule at 1530 every afternoon, operating a Dash 8-200 prop-driven 36-seater aircraft at $90 for the 55 minute scenic flight over the mountains. Your altitude on arrival at Ayacucho – 'Sanctuary of the soul' or more mournfully, 'Place of the Dead' – will be 2,750 metres/say 8,250 feet.

Arco de Triunfo, Ayacucho

Museum piece, Barrio Santa Ana

Dansa de las Tijeras Cultural Centre

Depiction of weaving and spinning

Pampa de Quinua

Sofas *Ayacuchenos*

Road to the Wari ruins

Battle Obelisk

Potted Profile. Since the **Battle of Ayacucho** in 1824 (see below), the Department of Ayacucho (population approaching 1,000,000) has lent its name to the principal city of the region, formerly solely known as Huamanga – 'the place where eagles are seen' (estimated population now 200,000). There are 33 churches in the municipality, mainly dating from the 1550's (hence 'Ciudad de la Iglesias'). The oldest (1540) and smallest (so tiny and seemingly neglected it is in danger of being obliterated) being that of San Cristóbal; to tantalise researchers and historians it is linked with the **Batalla de Chupas** of 1542. The immense Cathedral in the equally impressive Plaza de Armas was started in 1612 and completed in 1672. The city ('Ciudad Señorial') is noted for its rich traditional heritage, art and architecture, stemming right back from the time of the Wari Empire through to Colonial. The weather in Ayacucho is generally sunny and benign, as has been said of the inhabitants.

Accommodation Etc. Hotel Santa Rosa (Jr. Lima 166, tel.066.314614/315830), just around the corner from the Plaza, and originally the beautiful colonial Casona Gutierrez built in 1630, with a large flowery courtyard and fountain, is an excellent choice (cash only payment). All facilities, guest computer, and abundance of distinctive sofas ayacuchenas on the balconies, check-in conveniently at 0800 to accommodate the arrival of the overnight buses and the early morning Star Peru flights. Alternatively, lively ViaVia Café (tel.312834) is right on the plaza and very much in the heart of things, whilst immaculate sister hotel ViaVia 2, Alameda (tel.316041) is further out at 28 Julio Cdra 7. There are dozens of other choices, not least being Hotel San Francisco de Paula (tel.312353, Jr. Callao), complete with artefacts and mirador.

Amongst the plentiful excellent restaurants in Ayacucho, campestre 'Don Manuel' at Jr. Garcilazo de la Vega 370 is just one, offering comidas criollas and platas típicas such as cuy (guinea pig, of course), trucha, mondongo (soup based on cow's stomach and maize), chicharron (fried pork), **chuño** (sun-dried potatoes, requiring severe cold weather – as occurs at night on the altiplano – to complete the process; the basis for delicious soups) and puca picante (a red, spicy dish).

As for domestic matters, lavanderia 'Humanga Laundry' conveniently awaits you also in **Garcilazo de la Vega** (yes – the 16th century poet, son of a Spanish aristocrat and a royal Inca mother), at 265 and at Av.Ramón Castilla 518, with a third laundry at Maria Parado Bellido 215.

What to do and see. Possibly start with the magnificent gold leaf altars in the Cathedral on the eastern side of the spacious and flower-filled Plaza Mayor, floodlit at night. Immediately opposite in the arcaded courtyard of the former 1740 Colonial residence now housing the Prefectura justice department you will find the sombre prison cell wherein resistance heroine **Maria Parado de Bellido** was incarcerated by the Spanish prior to her execution in 1822. And working round the Square to the north in another fine casona is the Cultural Centre of the Universidad Nacional de San Cristóbal de Huamanga (UNSCH),venue for regular tradition shows and activities. And behind that is the architecturally impressive church of Santo Domingo dating from 1548.

The south western corner of the Plaza leads into traffic-free Jiron 28 de Julio full of interest and leading through the **Arco del Triunfo** commemorating the defeat of the Spanish. Just beyond is the colourful central market (Mercado de Abastos Carlos F.Vivanco) with the customary not-to-be-missed cornucopia of food and juice stalls, and an amazing abundance of local produce.

One mandatory excursion to undertake beyond the city is to the **Pampa de Ayacucho** where the Cordillera del Condor starts. Here (also referred to as the Pampa de Quinua) there is a 44m commemorative obelisk to commemorate the decisive battle on 9th December 1824 in which nationalist forces under **Mariscal Sucre** defeated royalist troops under the Viceroy of Spain, thereby confirming Peruvian independence which had previously been declared on 28th July 1821. It was at this point that **Simón de Bolívar** announced that henceforth Huamanga would be known as Ayacucho. The battle site lies some 35km from Ayacucho; combis are available, but an organised trip is better and easily (and economically) available. On the battlefield itself small boys offer a little tour for a few soles and are well-informed guides.

The picturesque small town of **Quinua** is situated just before the Pampa of that name. The traditional architecture has been attractively restored in response to the estimated 10,000 tourist who pass this way annually, and a range of Ayacucho menus are offered. There is also a museum in the building where The Act of Capitulation was signed by **Virrey La Serna**. Additionally, the town is famous for its pottery, with some 70% of the populace engaged on this activity – hence 'el pueblo de las manos que hablan'.

Along the same road at Km 22 are the important **Wari/Huari** archaeological ruins, the first known walled city in the Andes. Here there is a museum, and the remains of dwellings, mausoleums, public buildings, streets and the high perimeter defences may be determined. It is estimated that there were 50,000 inhabitants at Huari – much of the site is now overrun by prickly pear cactus. The Wari culture (successor to Los Warpas, Nazca and Tiahuanaco people) dominated in Peru during the period 500-1100 AD, pre-cursor to the Inca era. Ayacucho was the centre of this extensive Andean empire, known for the high quality of its ceramics, cloth, metal and stonework. Allow a full afternoon for the three strands (Pampa de Ayacucho, Quinua, Huari) of this trip.

Aside from visiting the many churches (e.g. San Francisco de Asís, Jr.28 de Julio cdra 3, and Santa Clara de Asís, Jr.Grau, cdra 3), three other recommended activities are as follows. Firstly, a visit to the very well presented **Museo Mariscal Andrés Cácere**s ('Wizard of the Andes' and twice President), complete with guided tour by the military at 28 de Julio cdra 5. Secondly, spend sombre and reflective time in the **Museo de Anfasep** dedicated to the memory of the Secuestrados, Detenidos y Desapericidos from the time of Sendero Luminoso (Abimael Guzman's Shining Path movement) and the recent 20 year civil war. Thirdly, **Barrio Santa Ana** is home to the Andina cultural artisan galleries and shops (including filigree silver and piedra de huamanga – the 'alabaster' scupltures), not least being the textile Museo Hipólito Unaune. Hence the accolade 'Capital del Arte Popular y de la Artensanía del Peru'. Also, the Mercado Artesanal Shosaku Nagase is a straight walk four blocks due north from the Plaza Mayor down 9 Diciembre, housing the Asociación Artesanos (outside the market in Parque Maria Parado de Bellido where a fine statue commemorates the resistance heroine, you will find a long stall laden with produce especial to the Andes, including grains, wines, liquors, and interestingly, coffee).

For its world famous Semana Santa celebrations ('Patrimonio Cultural de la Nación) all hotels and hostels in Ayacucho are fully booked well in advance, but there are many other carnivals and fiestas, featuring amongst other things the Andean speciality of firework towers arranged on tall bamboo frames with great creative ingenuity to ignite progressively producing spectacular effects (juegos artificiales/juegos pirotecnicos).

Moving on. You will be reluctant to depart, but if going back to Lima the 1100 daily Molina bus from Ayacucho's fine terminal terrestre Terrapuerto Libertadores is one way to go – no point in doing this scenic journey by night unless you really feel you must. Front seats, panorámica, upper deck, S/.70-90. Journey time 8hrs, but probably nearer to 2100 by the time you reach the final depot at Av.28 Julio through the Lima traffic. For the first 3hrs from Ayacucho the metalled road twists and turns through the mountains, and your altitude is still 4,736m/15,538ft around Santa Inés. Thereafter follows the long descent down the fertile valley to Chincha and Pisco (change here for Ica), ears popping as you pass avocado plantations and extensive commercial vineyards to reach sea level. The route joins the Pan Americana del Sur highway south of Cañete, from whence it is a straight run north along the coast to your destination.

Star Peru and LCP will of course fly you back to Lima, but if your ultimate goal is Cusco, then you may travel through the central highland towns of firstly Andahuaylas, and then Abancay, both of which well merit visiting for a day or two – or more. Read on.

Prefectura - 1740 Colonial residence wherein Maria Parado de Bellido was incarcerated 1822

Maria Parado's prison cell prior to execution (see text)

Vendor of locally produced coffee, Ayacucho

Casona Gutierrez 1630

Andahuaylas

Getting to Andahuaylas. This friendly and scenic Andean town can be reached by overnight bus direct from Lima, journey time approximately 16hrs (take the five o'clock bus – the earlier ones spend hours visiting various passenger collection points in the suburbs before leaving the capital). Cost around S/.80.00 By air, LCPeru fly daily, flight time 1hr 20min.

If travelling from Ayacucho, take Los Chankas buses (or Celtur) or use the morning minibuses, all from Terrapuerto Libertadores, the impressive Ayacucho terminus for a variety of destinations. Alternatively use any of the operators in Pasaje Cacéres for departures later in the day. From Ayacucho to Andahuaylas by bus and minibus (price S/.30-40) takes around 6hrs for the 270kms journey. A hired taxi for three or four passengers may be had for S/.60 per head and will complete the distance in 4hrs hard driving.

The entire way is paved now, cutting the previous 10-11hrs travel time by half, but the road twists and turns throughout (Gravol recommended), climbing first to the puna (altiplano) at 4,500m (13,500ft) where vicuña may be seen. Then it drops back to 2,500m, passing through patches of sweet corn, mandarin orchards, red quinoa and alfalfa, before rising to 3000m at journey's end. Many of the village names en route end in *bamba*, Aymara for pampas or plain.

Potted Profile. Undulating Andahuaylas lies in a scenic valley in the Department of Apurimac at a height of 2,980m/ 9,777ft. The population has doubled since the turn of the century to 70,000. The region was occupied by the Chankas and the Incas in pre-colonial times. The town, founded in 1533, is an important agricultural, mining and communication hub (phone code 083).

Accommodation Etc. There is a saying thus in these parts: *'No tenemos hoteles con cinco estrellas, pero tenemos un santuario en las montanas con milles de estrellas'*. In fact Andahuaylas has many good hotels, amongst which are El Encanto de Apurimac, Conquistador, Imperio Chanka, Sol de Oro, and El Encanto de Oro (at Av. Pedro Casafranca 424, tel.083-423066, very handy for Fitness Gym just up the hill at No.580).

Plaza Café D'Marce is popular, as is the Garabato Pub, especially for its incomparable *Té Piteado (Calientito)*, and for *Pasñawaccachi* – the 'drink which makes the girls cry' (not to mention *Chankaquicachi*). Restaurant Nuevo Horizonte at Jr.Constitución 426 specialises in vegetarian and health foods.

What to see and do. Festivals and Ferias need no excuse. The biggest of all is the ***Pukallay***, the famed Andahuaylas Carnival (last week of February/first week of March). As in other parts of the Andes, Christmas followed by New Year marks the start of some three months of serious festivities, terminating with Carnival (derivation: 'carne vale' – 'farewell to meat') and Easter, *Semana Santa.*

Look out for the phenomenon of ***Mallqui***: Eucalyptus saplings up to 20/25ft in height, well embedded in the ground and liberally laden with donated pots, pans, blankets, plastic buckets and even stools and little chairs, which are scaled by the brave encouraged by the local brass band until everything collapses amid great excitement. ***Negrillos*** are a common form of entertainment at these events – colourfully dressed tumblers and acrobats performing up and around an elevated platform or cross; not perhaps as wildly hazardous as older spectators will recall, but still spectacular. Yet another Andean speciality is ***Inkacha***, a dance with whips founded on the Inca heritage but later morphed into a form of silent cultural resistance to the harsh rule of *los Conquistadores.*

Other Attractions. Andahuaylas market (just off Av.Pedro Casafranca from 0630 Mon to Sat) contains the customary colourful array of everything, plus the usual highly economic food stalls. One highlight has to be the ***Rana Extracto*** vendor, whose speciality is live frogs in the blender to provide a cure for practically all known ailments and enhanced mental capacity. Over-ripe healing Noni fruit (imported from the Caribbean, originally from SE Asia) are added (with smell to match) together with maca powder from the root of a plant of the radish family grown in the mountains ('Peruvian ginseng'); the resultant mix is a potent cocktail, only for the brave.

The Sunday market (*Feria de los Domingos*) is the best part of a kilometre in length, and is packed with an astonishing assortment of local colour, culinary delights and hand crafts from outlying villages. For aficionados of embroidery, the fast disappearing skill of back-strap weaving (*tela de cintura*) may be located in Andahuaylas, whilst a visit through Gate 7 of the Municipal Stadium gives access to the renovated and recommended ***Museo Agroecológica/ Bioarqueológica.***

Near Sondor

Los Negrillos

Mallqui

Picturesque and rural ***Laguna de Pacucha*** is some 20kms/40 minutes from Andahuaylas and may be visited in minibus or taxi (S/.30.00, depending) en route to the Inca archaeological site of **Sóndor**, the approach road running along the shore of the enchanted bottomless lake. The preserved ruins and ancient terraces are full of evocative interest, and the summit at 3,300m/10,827ft affords splendid introspective views for miles and miles across the mountains. Also at the summit is an ***Intihuatana***, a stone obelisk (similarly found at Machu Picchu), translated loosely from the Quechua as 'the place where the sun gets tied' – in other words a granite solar clock. Cloaked in mystique, the shadow cast by the sun at the base of the rock serves as an astronomical calendar, indicating the changing seasons and in particular the arrival of the Winter Solstice (June in the southern hemisphere) and the moment for the important ***Inti Raymi*** (and the newly-introduced ***Sóndor Raymi***) rituals and celebrations. Decidedly worth the climb and not to be missed.

Moving on. You can of course return to Lima through Ayacucho, except having come this far, the only logical way is onwards in the direction of Cusco. You can do this in one hop on an overnight bus (10-12hrs for around S/.65) leaving at 1800/1830, but once again it seems a pity to miss the mountain scenery, and the chance to stay in the town of Abancay on the way. The road from Andahuaylas is fully paved throughout the 138kms and climbs up to 4000m/13,123ft at one stage, predictably sinuous and picturesque as it passes through a patchwork of intensive cultivation tended by the campesinos. A collectivo minibus costs S/.25, similar price to a bus; a taxi is more like S/.40 per head for the journey which takes around 4hrs.

Abancay

Arriving in Abancay. Travelling from Andahuaylas, the town of Abancay is first sighted from a distance of some 45km at the end of the distant glacial valley of the river Pachachaca, surrounded by beautiful green mountains. Not long afterwards it pays to stop to view the impressive old colonial **Pachachaca bridge** (signposted, 400m from the main road), and simultaneously to sample a glass of *cambray*, *refresco tipico* (much more palatable than the literal translation of 'spring onion' might suggest). Between the months of June and November the Pachachaca river is now a popular venue for white water rafting and canoeing, although Abancay itself sits on the banks of *el río* Mariño, tributary of the Pachachaca.

Potted Profile. Capital of the Department of Apurimac (total population approaching 200,000); Abancay population is 60,000. Height above sea level: 2377m/7800ft; phone code 083. For etymologists, the transliterated name flows from the Quechua word *amánkay*, which in turn is the amaryllis lily (*azucena*). Thanks to a micro-climate in the sheltered valley where it sits and despite the altitude, Abancay is home to bright bougainvillea, hibiscus, hollyhocks, cosmos, brugmansia, roses, gladioli and many other flowers reminiscent of a country garden – hence the claim 'City/Valley of Eternal Spring'.

The Battle of Abancay in 1537 was part of the confused in-fighting that beset the Conquistadors, but the town was not formally founded until 1575; in pre-Hispanic times it lay at the frontier between the zone of Inca influence and that of Los Chankas. Abancay sits strategically at the crossroads of the Inca highway between Nazca and Cusco, and that from Cusco to Pisco, through Ayacucho. Nowadays it is noted for mining and agriculture (notably sugar – and resultant *aguardiente* – and brandy).

Accommodation Etc. Comfortable Hotel Turistas is centred on a colonial style building at Av.Díaz Barcenas 500 (tel.321017) or you can try the Saywa hotel by the monument to Micaela Bastidas ('martyr for Peruvian Independence', wife of Túpac Amaru II, heroine, executed barbarically with her son in front of her husband – also killed – by the Spanish 1781). There is a plenitude of other accommodation options, eg El Peregrino Aparthotel, Av.Andres Caceres 390. La Leña restaurant (Sra Gladys – Prop., Jr. Libertad 116) is the highly popular venue for *pollo a la brasa* (golden rule: invariably accompanied by Inka Cola), or for vegetarian and health food go to Nuevo Horizonte (Jr. Constitución 426).

Plaza de Armas Abancay

Pachachaca Bridge

What to see and do. The Plaza with its customary cathedral is an attractive resting point, and whilst there is no museum in Abancay a recommended historical, close by, out-of-town excursion (taxi S/.7.00 – just 20mins – wait and return) is to the ***Casona Museo de Illanya*** located in one of the first colonial haciendas dating back to 1592 which in itself is of much interest, not to mention the 150 year old *caña* press and the collection of wari textiles. Best allow a couple of hours for this, all told. Carnival in late February/early March is renowned for the importance of its ethnic music.

Also just a few kilometres from Abancay is the archaeological site of Saywite, former Inca temple, easily accessed by taxi. Adjacent to Saywite are the thermal baths of Konoc (*Cconocc*), whose volcanic waters are much valued for the treatment and alleviation of arthritis, psoriasis and asthma; well worth visiting.

Equally from Abancay you may reach the remains of the key Inca city of **Choquequirao**, as yet scarcely 40% excavated but rivalling Machu Picchu in significance. From the Quechua/Aymara name of *Chuqi-K'iraw* is derived the literal translation 'Crib of gold', and from the assumed historical position of this extensive settlement comes the description 'Last Refuge of the Incas'. Geographically the site is situated on the truncated top of a mountain at 3,050m/10,010ft and is reached via a demanding 2-day trek on a 31km trail that rises and falls strenuously with the contours through the villages of Cachora-Chiquisca-Marampata. An alternative route lies through Huanipaca on a 15km approach trail, which takes 7 hours. Tours of 2/3 days are generally arranged by visitors through agents whilst in Cusco, but packages direct from Apu-rimak Tours (tel.083-321017) in Abancay Hotel Turistas equally include a guide and everything else you may need – tents, food and horses. Funding has apparently been provided for a 15 minute cable car to Choquequirao – certainly convenient but arguably at the expense of mystique and magic.

Moving on. The paved road out of Abancay to Cusco once again climbs to 4000m, up into the clouds, helpfully signposted as a *Zona de Neblina*, and yet another feat of civil engineering. The stunning views across the cultivated valleys to the far mountains compensate for the crawl of heavy vehicles obstructing the way, and soon after the little town of Curahuasi at 65kms (noted for its oustanding *anís* – aniseed) the road straightens. This enables the 198km journey to the Sacred City to be completed in little more than 4hrs. At 100kms look out for the spectacular suspension bridge over *el río* Apurimac and along the way keep an eye open for the elderberry (*saúco*) look-alike, the *arrayán* fruit (myrtle).

The journey to Cusco may be undertaken by taxi for around S/.200 for three passengers, but equally conveniently and comfortably, more economically (around S/.40 per head) and almost as expeditiously by bus from the busy Abancay terminal terrestre on Av.Pachacútec. As at terminals throughout Peru, the number of bus operators is nothing short of amazing – Expreso Sánchez, Turismo Cavassa, Los Chankas, Señor de Animas, Ampay, Bredde, Etecsa, Flores, Palomino, Tepsa, Movil, Wari, Cruz de Sur, Ferkaxi, Cromotex and so on – to mention but a handful – with differing permutations on fares and timings to Andahuaylas, Ayacucho, Cañete, Cusco, Ica, Nazca, Puerto Maldonado, Uripa etc etc. The one common thread is the necessity for travellers to pay boarding tax of S/.1.50 before departure (TAME: *Tame Municipal Embarque*).

Abancay to Lima by bus is straightforward, with many options on price and timing. One such is with Flores leaving at 1400 daily down the 'springtime valley' Apurimac and then climbing out over the pass at an estimated height approaching 5,000m/16,000ft, across the remote and uninhabited altiplano before descending and passing through Chalhuanca, following down the long fertile valley from Puquio all the way to Nazca at kilometre 497 through the productive vineyards. Then past Ica on '*La Pan Norte*' to reach Lima (870km) any time from 0600 onwards the next morning (17hrs upwards). All for S/.80.00 – same as the price of a seat from Cusco, where the bus originates. As always, go for daylight travel as much as possible.

Apurimac *campesinos*
Road through the mountains
The Apurimac agricultural tapestry

Cusco

Arriving in Cusco. The busy airport at Quispiquilla (less than 2km from the city centre) receives over 30 flights a day from Lima (55mins duration). Principal operators are LAN (16 flights), Avianca, Star Peru and Peruvian Airlines. Intended extension of the runway to accept international flights is controversial, not just for Cusco itself (already submerging under a torrent of well over one million visitors a year and in danger of losing its identity), but also for Lima and the remainder of Peru. Air travel is generally swift and convenient, but keep in mind the options for arriving by road – from Puno (5/6hrs), Andahuaylas (10/12hrs), Abancay (4½hrs), and even Lima (20/24hrs). Subject of course to landslips and flooding in the months of January to April.

Profile. The Imperial City of the Incas, generally accepted to date from 1100AD, needs no introduction and little amplification to information readily available elsewhere. Population half a million, height 3310m/10,860ft (considerably higher than Machu Picchu which is 2380m/7808ft). Phone code 084.

Templo y Convento de la Merced

Sacred Valley across the rooftops

Museo Histórico

Cusqueñan dress

Accommodation. For a treat, certainly go for El Monasterio (Calle Palacios 136, tel.604000) or any of the competing top end rivals. One well recommended mid-range choice is Andenes al Cielo (Choquechaca 176, tel.222237) based in an old historic home, roof terrace, guest computer, evening log fire. However, the beautiful little garden and views, plus the welcome at Hostal El Balcón (Calle Tambo de Montero 222, tel.236738 – 10 min walk up from the Plaza) have the edge. Oxygen is not normally provided in hotel rooms, nor necessary. Restaurants (and more mundanely, handy lavanderias), are on every corner in Cusco. Two coffee houses hidden in Calle Espaderos just off the Plaza at nos. 116 &120 respectively are Super Cafe Extra and Cafe y Chocolate Confiterla.

THE INCA TRAIL

(With grateful thanks to contributor Henry Morlock)

Trekking to Machu Picchu on the Inca Trail is included in my top ten of achievements and experiences so far in my life. The combination of awe-inspiring mountain scenery, lush cloud-forest, subtropical jungle and the Inca sites passed along the way culminating in the famous archaeological site of Machu Picchu are all contributing factors.

My old school friend Lucy and I planned and dreamt about our trip for months. Arriving in Cusco to adjust to the high altitude for a few days we loved the old city with its carved wooden balconies and Inca ruins. With a mix of anticipation and excitement we finally met our group of 10, in a painfully early start. It was no time for self-doubt on whether we could do this...we set off.

For three days we walked in what felt like a mainly uphill direction. We heard the Urubama River roaring below us and climbed the ominously named Dead Women's Pass, all the while surrounded by the most dramatic scenery and formidable snow capped peaks.

The final day we arrived at the Sun Gate in time for sunrise. I remember finding a quiet rock to sit down on and watch the sun come up. I will never forget that sense of elation as the mist began to clear revealing the ancient ruins of Machu Picchu. It was a magical and unforgettable sight. We had done it! And not only that, we had managed to enjoy ourselves too.

What else to see and do. It is taken as a given that you will go to Machu Picchu, and even undertake the challenge of The Inca Trail (an experience succinctly encapsulated in the foregoing first hand account by Henry Morlock). But do bear in mind the wise words of Miguel Gongora Meza: '*Avoid spending all your time behind the lens of a camera. Immerse yourself in the beauty and atmosphere of this spectacular site, the zenith of a civilisation that placed the needs of its people and mother earth Pachamama above all else... Machu Picchu is not for selfies*'. You will also probably find your way to all or some of the following: Ollantaytambo; Tambomachay; Pisac; Tipon; Moray; Chinchero; Saqsayhuaman; Q'Enqo; Puka Pukara and Pikillaqta.

What follows addresses additional activities in Cusco that otherwise may or may not have escaped your attention. The daily **Mercado Central San Pedro** (Av. Santa Clara) is often missed; when you get there try *Tocosh Andino* made from fermented potato peel. And whilst in the vicinity be sure to call in at Illari (Av.Santa Clara 419) for your souvenir shopping. Strolling back to the Plaza you will pass the impressive **Templo y Convento de la Merced de Cusco**; be certain to stop and enter the beautiful arcaded interior and priceless artefacts within. And if visiting the Cathedral on the Plaza, remember to take your passport (copies not accepted), otherwise they won't give you the audio tour cassette which is included in your entrance fee.

El Baratello is the Sunday Market, good for craft work. And if back-strap weaving is still eluding you, then go to The Center for Traditional Textiles, Av. El Sol 603 (just up from the LAN and Peruvian Airline offices at 627/627A). And continuing the cultural theme, the ***Centro Qosqo de Arte Nativo*** at Av. Sol 872 has a wonderful show of dress and dance at 1900 nightly, seven days a week for S/.25.00, all too often missed by visitors.

The months for bullfights in Cusco are July and November. And as a finale to your visit, spend time at the uniquely informative **Potato Museum** and Restaurant at Calle Cascaparo 116 (tel.252131), bearing in mind that the now ubiquitous potato originated in southern Peru and has been cultivated for some 10,000 years, currently resulting in over 1,000 different kinds. It was introduced in Europe by the Spanish around 1550. Annually on 24th June the farmers in the Andes have for centuries viewed the stars reflected in water in dedicated pottery plates to obtain a forecast for the impending rainy season three to eight months ahead. Should the night stars appear dim, this indicates the rains will be late, and the farmers delay the planting of their potatoes. The latter-day scientific explanation of this is the presence (or otherwise) of cirrus cloud which often heralds the onset of *El Niño*.

Moving on. When the time comes to leave, out of all the options available (such as the many return flights to Lima – and elsewhere), give a thought to taking the scenic bus to Puno (0830/1500 and 1000/1700), or use one of the manifold regular services, from whence Lake Titicaca and even La Paz await... (as detailed in Section I earlier).

Convento de la Merced

Aguardiente vendor

Centro Arte Nativo.

And even if you arrived by bus, remember that reverse journeys always look different and produce new vistas – such as the beautiful valleys on the way back to Abancay, with their frost-free micro-climates and scarlet bougainvillea and plantations of papaya. An easy road, 190km, four and a half hours; superb scenery across and through the mountains; the route follows *el río* Blanco, favoured for white water rafting these days. You will share the route with Bredde, Wari and Flores buses; approaching the final pass at 4,800m/15,748ft with 40km to run you will pass the Saywite Community and archaeological site, and then catch your first glimpse of Abancay to the west in the glow of the setting sun.

However, and all that aside, the time has now come to visit the Northern Highlands of Peru (see following Section III). But just before you do that, here follows a short treatise on the spectacular and colourful Andean Scissor Dance which is to be found centred on the Departments of Apurimac, Ayacucho and Huancavelica.

Cusco line steam locomotive (made in Edinburgh) preserved in *Parque Urpicha* adjacent to *Terminal Terrestre*.

Ritual Danza de las Tijeras

Introduction. This is a spectacular ancient symbolic and ceremonial dance from the heart of the central Peruvian Andes. The colourfully attired dancers manipulate in one hand with great skill and timing two metal blades as they perform, to sound a percussive rhythm to accompany the traditional background musicians playing the violin and the inverted harp of the region. It is these two mesmerically clinking and flashing steel blades that give rise to the description of the ritual as '**The Scissor Dance**'

The performance. This customarily comprises two competing dancers, vying one against the other in giving alternating displays of increasingly complex physical and athletic content. However, the established sequences are often followed equally dramatically by a solo scissor dancer. In its finality, the end of the ritual may on occasions extend to frog-swallowing and toad-tasting, cheek- piercing needles and to other extremes.

History & Symbolism. Rooted in the mystical spirituality and pre-Columbian beliefs of the dwellers of the Andean high mountains, this ancestral dance owes it origins to a blending of sacred ceremonies, not least being obeisance to the deities and spirits of the peaks (the *Apus* and the *Wamani*). Also to the Sun and Moon, and to Mother Earth (*Pachamama*), the harvest and fertility. Subject to suppression in colonial times because of its links to the Incas and earlier, the dance survived as a manifestation of cultural identity and defiance. Today it is a proud symbol of Peru's artistic inheritance.

Summary. In Quechua the phases of The Scissor Dance may be summarised as follows. ***Pagapu*** – the solemn offerings to *Pachamama* and the *Wamani* and *Apus*. This is followed by ***Pacha Tinkay*** and ***Pachakuy*** – the dedication and donning of the costume. And finally ***Atipanakuy***: expression and challenge of the dance, demanding courage and great agility, and years of practice and preparation. The Scissor Dance is in the UNESCO Representative List of the Intangible Cultural Heritage of Humanity.

Andean Textiles and Designs

Section III
The Northern Highlands

"No matter which way the wind howls
The mountains will never bow to it"

Huaraz, *Cordillera Blanca* & Huánuco

Huaraz is the recognised principal entry point for the legendary ***Cordillera Blanca*** – the spectacular mountain range within the vast mountain range that is the Andes. Just over 400km from Lima, the bus journey by road is served by all the principal operators, including Cruz del Sur, and takes 7 to 8 hours in return for between S/.60-S/.90. The 65min flight from Lima to Huaraz Airport is undertaken by LC Peru thrice weekly on Tuesdays, Thursdays and Saturdays. The city has largely recovered from the cataclysmic earthquake of 1970 measuring 7.9 on the Richter scale that took the uncounted lives of perhaps 150,000 in the area (including those of the inundated and obliterated adjacent township of Yungay, one-time centre for cut-flower rose growing). Estimated population today is 150,000, height 3,052m/10,013ft (phone code 043); possible etymology: Quechua *waraq*=sunrise.

As usual, **accommodation** options are many, including: *El Patio*, Av.Monterrey (tel.424965), colonial with garden; *Hostal Colomba*, 210 Francisco de Zela (far side of *Río Quilcay*)(tel.421501), hacienda, gym; *Hatun Wasi*, Jr.Daniel Villayzán 268 (tel.425055), with mountain views from the roof terrace. When the time comes to leave, regular services are available to **Huallanca** providing an interesting Andean transit, and from there onwards to **Huánuco** (below).

Huaraz is a mecca for climbers and trekkers visiting the ***Parque Nacional Huascarán***, which embraces the *Cordillera Blanca*. From the city itself the snow-covered twin peaks of Huascarán (at 6,768m/22,205ft the highest points in Peru) may be seen, and the summit of the adjacent Mount Huandoy, plus reputedly 23 other peaks. It is also said that on the higher of the two Huascarán peaks, the force of gravity is the lowest to be recorded on earth… The National Park is regulated by strict conditions of entry and is a UNESCO natural world heritage site. The area was peopled long, long before the Spanish 'founded' the city in 1574, and there are many important archaeological sites in the vicinity, including the pre-Inca fortified temple of ***Chavín de Huantar*** (800BC), relic of the Chavín people and another UNESCO world heritage site. It is linked with the ***Museo Nacional Chavín*** (both closed on Mondays) just 1-2kms from town.

Amongst many other attractions, from Huaraz the ***Pastoruri Glacier*** is readily accessible. Until 2007 it was possible to walk on the ice, but erosion accelerated by global warming led to cessation of this facility. In 2011 innovative tours to the glacier locale re-opened with emphasis on climate change awareness; as for other glaciers in Peru (and elsewhere in the world) the *Pastoruri Glacier* will have disappeared in a matter of years.

Huánuco. *La ciudad de Caballeros de León Huánuco* (1,913m/6,275ft; phone code 062) is on the site of the former Inca town of Yarowilca, 're-founded' by the Spanish in 1539. It sits at the confluence of *el río* Higueras and the dominant river Huallaga (which flows onwards past Tarapoto to join the Marañon beyond Yurimaguas to meet the Ucayali at Nauta, to become the Amazon through Iquitos and thence westwards all the way to Colombia and Brasil). Equally strategically, the city straddles the Lima-Tingo Maria-Pucallpa highway and is the capital of the Department of Huánuco, wherein lies the *Cordillera Huayhuash* which includes Yerupaja (6,617m, the second highest peak in Peru), Siulá (6,356m) and Ronday (5,870m).

Interestingly (or frighteningly) the population of Huánuco has expanded exponentially in the past decade (in common with other cities in Peru – and worldwide) viz: 75,000 in 2007; 173,000 in 2014; 200,000 estimated in 2017. Geographically, Huánuco is in the *yungas* (see Glossary) eco-zone of the eastern slopes of the central Andes. It was the birthplace of President (and General) Mariano Ignacio Prado in 1825, and (who knew?) of Daniel Alomía Robles (1871), composer of *El Cóndor Pasa.* The city *Plaza* is on a grand scale, with its modern cathedral may. ***El Templo de las Manos Cruzadas*** is part of the archaeological ruins dating from 2,300BC to be found at **Kotosh,** on the left bank of the river Higueras just 4kms from Huánuco. Slightly further on at 6kms and adjacent to the first ruins is **Quilla Rumi**, noted for its *arte rupestre*, the pre-historic 'rock art' to be found in caves, grottos and caverns. The *pueblito* of **Huacar** 30kms out of town is renowned as a 'time-warp' Andean village, whilst 2½hrs ultra-scenic drive away on the main highway (one third of the way to Pucallpa) brings you down to **Tingo Maria,** at 655m in *la selva alta* with its coca, coffee, sugar cane, tea, cocoa and even rubber. Population 40,000 and known as *La Bella Durmiente* (Sleeping Beauty) on account of the topography and silhouette of the overlooking mountain (one of the 'seven marvels of Peru). Much to enjoy here, albeit of the Amazon rather than the Andes, including *platos tipicos.* For accommodation, on the outskirts of town is *Madera Verde* (Av.Universitaria, tel. (062)561800) with two pools, cabins, butterfly farm and adjacent wildlife rescue centre; 10 minutes away on the Carretera a Castillo Grande (km3.5) is *Albergue Ecológico Villa Jennifer* with a little zoo, pool, birdlife, flowers (tel.062.603509).

Accommodation in **Huánuco** ranges from the Grand Hotel with gym, pool and restaurant (775 Jr.D.Beraún, tel.514222), to well recommended *Las Vegas* (28 de Julio 940 on the *Plaza*, tel.512315). And when the time comes to move on, Huánuco airport is served by two LC Peru flights from/to Lima daily, plus day and night buses for the 8hr road trip to/from Lima (including Cerro de Pasco, Junin and La Oroya en route; plus access to/from Huaraz as above – and eastwards to *la selva.*

Symbolic representation of an Andean god - Chavin

El Señor de Sipan medallion - Chiclayo

Andean handwork

Hotel *Puerto Mirador,* Moyabamba
Entrance to *Orquideario Waqanki* (and Yosi, your guide)
Work on the *chacra*
Taquisho Mirador
Cloud forest
View from the *Mirador*
River Mayo valley

Moyabamba

Getting to Moyabamba. Heading from the coast from Chiclayo, accessible by bus through Chachapoyas en route to Tarapoto, journey time 12hrs and more. Alternatively, coming from the east, fly in to Tarapoto (or arrive there by road via the cloud forest from Yurimaguas) and stay in *La Ciudad de las Palmeras* at resort Puerto Palmeras, or in town at *La Patarashca* or family-run *El Mirador* (in descending order of economy, but all good). From there, a shared car (4 pax) to Moyabamba (e.g. from Empresa Etrisa at Jr. Alfonso Ugarte 1096) costs S/.20 (or a nearby combi for S/.12); around 100km, under 3hrs, ascending from 350m/1,100ft in Tarapoto to 860m/2,700ft in Moyobamba climbing the eastern facing slopes of the Andes, where are situated the *yungas* (see glossary).

Profile. *Ciudad de las Orquídeas*, agreeable capital of the Department of San Martin, population 86,000, phone code 042. Founded in 1542 by the Spanish (their first 'jungle' city) on the site of a former Inca settlement, but recent earthquakes (1990/91) obliterated the last remaining vestiges of colonial buildings. Located in the *Río Mayo* valley and on the surrounding slopes, one of the coffee growing and processing regions of Peru. The Moyabamba *Festival de la Orquídea* takes place at the end of October/beginning of November.

Accommodation Etc. *Puerto Mirador* (Jr.Sucre, tel: 562050) is away from the centre in a marvellous setting looking across the valley to the hills beyond with birdsong from dawn to dusk; large circular pool, restaurant terrace with a chef equal to the printed menu, individual cabins in the gardens. More centrally on Jr. San Martin are mid-range hotels *El Portón* and *Orquídea del Mayo*.

What to do and see. Continuing the orchid theme, visit ***Orquidear io Waqanki***, 3km out of sprawling Moyobamba, but well worth it (S/.10.00) (especially if Yosi is your guide). Here you will see many varieties of orchids in their natural habitat (not great banks of blooms as at Chelsea Flower Show), and having done that, climb the *mirador* for a view of the city and an unsurpassed concentration of *colibri* humming birds (take repellent if you are susceptible to mosquitos). Just across the road are the ***Baños Termales San Mateo***, very well done and highly popular (S/.3.00 adults; S/.1.50 children/*ancianos*). Back in town, a visit to ***Taquisho Mirador*** is essential, with wonderful views over ***Puerto Tahuisco*** across the river Mayo valley, birds galore (and adjacent to the Chocolate Museum). An excursion may also be made to ***Laguna Pomacocha***

Moving on. The *Terminal Terrestre de Moyobamba* at Av.Grau cdra 6 is the staging post for long distance coaches from Tarapoto and Chiclayo (as above), and also those for Cajarmarca, Piura, Jaén, Tingo María, and even Ecuador. Immediately opposite the terminal across the dual carriageway at Grau 640 are two firms (*Cruz Hermanos* and *A Ser de Díos*) running minibuses throughout the day to Chachapoyas (S/.20.00 and 5hrs); book early for the prized front seat.

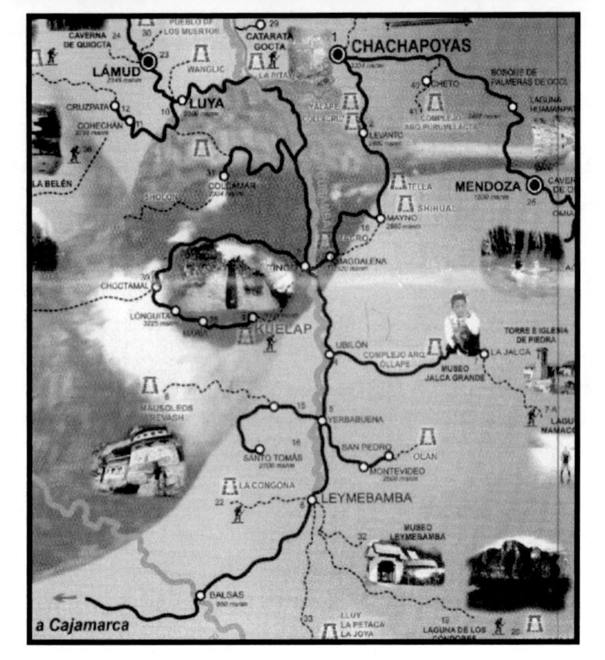

Chachapoyas

Getting to Chachapoyas. As above from Moyobamba/Tarapoto, or by 12hr bus from Cajarmarca. Particularly when coming from the semi-tropical east, be prepared for the much colder, fresher climate of Chachapoyas at an altitude of 2,350m/7,500ft. There is an airport for light aircraft, but no regular commercial services (the runway has a sharp drop at either end), although military *Grupo Aéreo 42* fly passengers from both Trujillo and Iquitos.

Profile. The somewhat isolated city of Chachapoyas was founded by the Spanish in 1538, albeit on or near the sites of earlier people. These were most notably the *Chachapoya,* the little-documented 'Warriors of the Clouds' of the northern regions of the Andes, latterly subjugated by the Incas and subsequently the Spanish. One etymological explanation for the city name is the Quechua word '*sachapuyos*' translating as 'mount of haze'. The Plaza is noteworthy for measuring precisely 100metres square. Population today is estimated at 35,000; phone code 041.

Accommodation Etc. There is a *Casa Andina Classic* a little way out of town, whilst directly on the Plaza are the *Revash* (tel.477391) and the *Belén* (tel.47830), both attractive, traditional and economic. *La Casona de Chachapoyas* (until recently named *Casa Vieja*) one block from the Plaza at Chincha Alta 569 (tel: 041.477353), has distinctively individual rooms around a courtyard and generally an open fire in the drawing room (Café *Terramia* is just along the street at No.557). There are of course many other options, including *Hostal Amazonas* (Jr.Grau 565, tel.478839).

What to do and see. Within relatively easy distance of Chachapoyas are a host of archaeological and historic sites, and of course the city is one starting point for visiting the spectacular pre-Inca walled fortress of ***Kuélap*** (see below) and also the ***Gocta*** **Falls** (one of the highest in the world). Day tours are easily arranged through operators on the Plaza and elsewhere. Nearer to hand, there are two small museums in Chachapoyas (*Ministerio de Cultura* at Ayacucho 904, and *Santa Ana* at Jr.Santa Ana 1054) to be visited, together with the cathedral and its modern interior. The battle of *Higos Urco* took place nearby on 6th June 1821, part of the struggle leading to the declaration of Independence on 28th July 1821, and this significant victory is commemorated by a monument in Independence Square.

Moving on. Buses may be had in all directions, namely Chiclayo, Trujillo, Lima, Cajamarca, and also to more local destinations. For **Leymebamba**, the *Terminal Terrestre* (taxi S/.3.00 from Plaza de Armas) is also the place for your combi to undertake the 2-3hr/83km journey on paved roads for S/.10.00 using Raymi Express, Amazonas Express or Karlita. The route, full of rural colour, follows the course of the Utcubamba river, passing the townships of Tingo (where there is a fork in the direction of *Kuélap*) and Yerbabuena (extensive district market every Sunday).

Running north west out of Chachapoyas the road drops down and once again picks up the Utcubamba river to reach **Bagua Grande** (440m/1,444ft). From there you may continue firstly to **Jaén** in *la selva alta* (729m/2,392ft; pop.150,000), and then north to **San Ignacio de la Frontera** (the clue is in the name) where you will find more Peruvian coffee, honey and grenadines, and an important sanctuary for the spectacled bear. Total journey time by bus is about 4½ hours for approximately 250kms. The town is quite small, population around 15,000, and a few miles further on lies Namballe, just short of the border crossing with Ecuador. Alternatively, tracking west after Bagua Grande will eventually lead to Piura, the oilfields and the Pacific seaboard; buses ply these routes.

(La Casona de Chachapoyas)

(Direct delivery - Raymi Express)

Kuélap

Getting to *Kuélap*. Tour operators in Chachapoyas run trips to the citadel daily (as they do for the Gocta Falls), price per head up to S/.60.00. Before the Presidential inauguration of the *Kuélap* cable car at the end of January 2017, transit of the 38kms dirt road from Tingo (approx 1hr from Chachapoyas) up through Choctámal took as much as 2hrs each way. It was not without scenic interest (and for most of the time it appeared you were ascending the wrong mountain), but the advent of the *Telerifico*, built by a French-led international consortium using helicopters and mules in the course of a year, has reduced the access time to 20 minutes. Each suspended cabin takes ten passengers, price S/.20.00.

Likewise from Leymebamba, pre-cable car the total journey time was 2½hrs each way and an all-day private car, wait and return, cost S/.200.00. Now it is under an hour to the cable car just above Tingo, and collection at the end of the day may be pre-arranged.

The Mountain Fortress. Towering and formidable *Cuídad Chachapoya* is massive, comprising huge amounts of rock. It was built continuously by the Chachapoya people over the period from 500AD to 1100AD. It consisted of an estimated 500 protected 'roundhouses' occupied by 3,000 to 4,000 people, and covered an area of 6 hectares (15 acres), running for over a kilometre along the mountain crest at a height of 3,000metres/nearly 10,000ft. 1200AD saw the rise of the Incan Empire, confrontation and partial subjugation of the Chachapoya. *Kuélap* was ransacked by the invading Conquistadors in the sixteenth century, only to be 're-discovered'.in 1843.

In recent years some sensitive reconstruction and restoration has been undertaken where necessary, but for the greater part the ruins and structures are largely untouched, including the cloud forest vegetation of orchids, ferns and small trees. From the reception point and car park there is a one kilometre uphill track to the fortress, and the tour can be something of a scramble in places (definitely flat shoes/trainers). Access can be made on horseback if preferred – S/.10.00 for a nimble mount. Site entry is S/.20.00 (half price for *ancianos*, S/.2.00 for children. An official guide (indispensable) costs S/.40.*00*. Considerably less congested than Machu Picchu, a visit to windswept *Kuélap* commanding views far and wide across the distant mountains with a big sky of scudding cumulo-nimbus clouds offers the perfect opportunity for solitary contemplation of how life must have been in the citadel in those far off days of the first millennium.

Leymebamba Plaza

Kentitambo

Humming Birds

Museo Leymebamba

View over Leymebamba

La Casona

Leymebamba

Getting to Leymebamba. As above, use Raymi Express from Chachapoyas. Coming from the other direction, an early morning bus from Cajamarca will pass Leymebamba in the early afternoon.

Profile. Essential reading for visitors is ***Los Chachapoya y La Laguna de los Cóndores*** (2016) by Adriana von Hagen (see below), which contains not only a wealth of information about the past inhabitants of the region but also their present-day successors. Leymebamba, re-founded by the Spanish in the valley at the meeting point of the rivers Atuén and Pomacocha, still retains its charm as an agricultural and traditional community of perhaps 6,000 people, connected by road to other population centres for hardly more than fifty years. Altitude 2250m/7,382ft; phone code 041.

Accommodation Etc. Conveniently in the town centre is friendly and helpful **La Casona** just off the Plaza at Jr.Amazonas 223 (tel.630301), with an old courtyard full of greenery and a surrounding balcony. **Kentitambo** is a luxurious lodge comprising five chalets set on a beautiful tree-covered slope some 10 minutes motor-taxi ride out of town. The flower-filled grounds abound with hummingbirds; the hospitality lacks for nothing, including a cosy log fire after dinner in the main house. Not to be missed on any account: it is the home of Dr Adriana von Hagen, noted authority on the Chachapoya, prime-mover in the salvage and recovery of more than 200 mummies and associated burial offerings from their vandalised site at the nearby *Laguna de los Cóndores.*

What to do and see. Rural Leymebamba is rewarding to visit in itself, but as explained, it is also a good starting point for going to ***Kuélap***. Additional tours may be undertaken to **Chachapoyan sites** within one-day reach at *La Congona, El Molinete* and *Cataneo.* The trek (and return) to the important *Laguna de los Cóndores* archaeological site, walking and riding, requires three days. Closer to hand, immediately opposite Kentitambo (see above) is ***Museo Leymebamba*** ("the best little museum in Peru"), home to the mummies, ceramics, textiles and other artefacts rescued from the looters bent on destroying and robbing the *Laguna* historic burial *chullpas*. The museum is open every day (except Christmas and New Year) 1000-1630 or by prior arrangement. The Sunday district market at **Yerbabuena** just 20 minutes down the road from Leymebamba is exceptionally colourful and varied, whilst further up the road in the opposite direction and beyond Kentitambo, condors may be sighted.

Moving on. Combi Karlita will pick you up from the top corner of the Plaza any day at 0730 (the local agent is in the general store two doors away); the fare to Cajamarca is S/.30.00, distance 260km approx., journey time 8-9hrs. Buses from Chachapoyas pass the Leymebamba Plaza just after 0800 for a similar fare. The first 5-6hrs of the journey are spent memorably en route to Celendin: the road is very narrow and sinuous, twisting and climbing in and out of the clouds, and is exceptionally spectacular, even by Andean standards. After 57kms you drop down to the township of Balsas at 960m/3,160ft in *la selva alta* to cross the Marañon River by way of the Chacanto bridge before climbing back into the mountains again to reach Celendin (founded 1802, pop.26,000; phone code 076) at 2,645m/8,678ft. Thereafter the 107km journey is more smoothly but less dramatically executed on a well-engineered major highway in 2-3hrs (car/bus), to reach Cajamarca just in time for tea.

The new cable car from Tingo to *Kuélap*

Fellow travellers

Angel's Trumpets on the road from Leymebamba

Plaza de Armas, Cajamarca

Cajamarca

Arriving in Cajamarca. There are four flights into Cajamarca airport daily from Lima, two by LC Peru and two by LATAM (LAN+TAM = "together further"); flight time scarcely more than an hour. Regular buses run direct from Lima (870km/advertised as 15hrs) including luxury Cruz del Sur and *LiNea.* Or travel from Trujillo (7hrs)(perhaps after spending time at the resort of Huanchaco), or from Chiclayo (6hrs). Scenically as above when coming from the direction of Chachapoyas.

Profile. Population 225,000, phone code 076, height 2750m/9,000+ft (higher than Chachapoyas yet somehow less bleak), Cajamarca, capital of the eponymous province and department (the 'Little Switzerland of Peru'), lies in the valley of the river Mashcon on a site that has been peopled by successive cultures for over 2,000 years, including during *el imperio inca.* It was at Cajamarca that treacherous Conquistador Pizarro 'ambushed' Emperor Atahualpa in 1832 (Battle & Massacre of Cajamarca), leading to Atahualpa's capture, imprisonment (plus payment of a huge ransom in gold) and subsequent murder by garrotting in the *Plaza Mayor.*

Nowadays the city is the commercial and touristic centre for the northern highlands, ranked 13th in size in Peru. The goldmine enterprise at Yanacocha has had a major impact on the economy and local development in recent years, not all positive, especially environmental. Unchanged is the production of local cheese, yoghurts and *manjar blanco* (*dulce de leche* with cinnamon and any number of other hidden additions).

Accommodation Etc. If your bank manager agrees, by all means stay at shiny Costa del Sol directly on the Plaza (tel.076.362472) or for great ambience at the excellent Laguna Seca (tel.076.584300), out of town near *Los Baños del Inca.* At 3-star level, Hotel Cajamarca (tel.076.362532) is ideal, comfortable and friendly, set around a colonial courtyard with a restaurant, 2-mins off the Plaza (Dos de Mayo 311). For a longer stay, go to recently opened *Hatuchay Inka Apart* Hotel (Sra Verónica Armas Moreno, tel.076.357059) just up the hill at Dos de Mayo 221. Two sweet-toothed cafés to try: *Heladería Holanda* on the Plaza, and further down Amalia Puga at 554, *Cascanuez* (with handy laundry *Dandy* No.545 directly opposite).

What to do and see. The ever-burgeoning outskirts of Cajamarca come as an unwelcome shock after the tranquillity of Leymebamba and the Utcubamba valley. However and happily, the *Plaza de Armas* and environs have retained

Traditional Costumes

Cajamarca

their calm and dignity of 20 years past, and of the Colonial era. The traffic in the city appears to be perpetually on the point of gridlock; for peace of mind, walk whenever you can.

Conveniently there are major points of interest on the flower-filled main square with its greenery and three-centuries old fountain, and within a stone's throw, starting with the 1776 **Cathedral**, and the **Bishop's Palace** across the road where the money changers operate. Facing them are the almost four-hundred year old **San Francisco Church** and the adjacent ***Museo de Arte Colonial***. Just 100yds further along Amalia Puga street you will find Atahualpa's prison, the ***Cuarto de Rescate***, and there is colonial architecture to be seen in all directions. As always, the city *Mercado* is a colourful place to visit and sample, but for ***Productores Artesanales*** visit *El Rescate* at Jr.Comercio 1029 (*Plaza* top corner, 100m past the Tourist Police) or *Feria Artesanal Los Frailones*, at Jr. Dos de Mayo 225. And for the Cajamarcan dairy products and pastries, speciality shops abound, with a concentration at the top end of Dos de Mayo.

There is much to do in and around Cajamarca, and should you have a day or two in hand then this is the perfect place to procure a bespoke pair of handmade shoes. Three further museums not yet mentioned are: the **Archaeological and Ethnological Museum,** ***Museo Silvo-agropecuario*** and ***Museo Arqueológico Horacio H Urteaga.*** Just a few kilometres out of town are the hot springs, the well run ***Los Baños del Inca***, whilst slightly further afield at 8kms are ***Las Ventanillas de Otuzco***, the macabre 'hanging tombs' burial niches of the Chachapoya people cut high in the cliff face. You may also visit the little Andean village of **Llacanora** and then go high into the mountain pine forests, the roadside purple with wild lupins, and look down on 'the green valleys dotted with dairy herds far below, where the concepts of the infinite and the eternal become realities'.

Moving on. Logically the reverse of arriving, in whatever direction. No daytime buses to Lima; evening departures 1800/1830, theoretically arriving in the capital at 0900/0930. However, the advertised travel time of 15hrs can easily stretch to 17hrs when the *Pan Norte* is under repair (forever) and the city traffic heavy (always). Cost is between S/.90 and S/.140.00. Cruz del Sur are on Atahualpa at 606 and Li*N*ea at Atahualpa 318 (also serving Trujillo and Chiclayo).

Now you are ready to pack, head to the airport and start your adventures in the wonderful mountains of Peru!

Glossary

Apacheta : cairn of stones built by travellers to ensure safe journey and good fortune.
Apu : (Quechua) two meanings – mountain, and 'Spirit or protecting Deity'.
Arandano : Blueberry.
Callana : earthenware Andean enclosed cooking receptacle, two little pottery handles at the sides, 'mouth' for access, used particularly for roasting corn (*maiz tostada/cancha*).
Campestre : literally 'in the countryside'; outside, in the open.
Cancha : roast maize.
Carrizo : bamboo firework frames (see also *juegos* below).
Cau cau : tripe, cow's edible stomach lining.
Chicharron : fried pork.
Chullpa : a burial chamber, generally associated with the Chachapoya people.
Departamentos : Peru is comprised of 24 Departments, formed in turn of Provinces.
Inti : Sun
Intihuatana : a stone obelisk (may be seen at Sóndor and Machu Picchu), translated loosely from the Quechua as 'the place where the sun gets tied' – in other words a granite solar clock. Cloaked in mystique, the shadow cast by the sun at the base of the rock serves as an astronomical calendar, indicating the changing seasons and in particular the arrival of the Winter Solstice (June in the southern hemisphere) and the moment for the important ***Inti Raymi*** (Sun Festival).
Juegos artificiales/juegos pirotechnicos : elaborate firework arrangement on bamboo frame, igniting sequentially.
Locro : mashed confection of pumpkin, potato and *habas* (broad beans); Andean ingredients for a dish found throughout Peru, served with rice.
Loncco : formerly a term applied to *Arequipeñan campesinos* (country people) descended from the first Spanish settlers of 1540. Largely history now, but *loncco* expressions are still in use in Arequipa, amounting to a hidden language.
Mondongo : soup made with cow's stomach and maize.
Páramo : inter-tropical mountain ecosystem, generally found from 2,700m to 4,000m (*el páramo andino).*
Puquina : together with Aymara and Quechua, one of the three languages of the Incan Empire, once spoken by ethnic groups in the vicinity of Lake Titicaca. Extinct now, but remnants of *Puquina* may still be encountered in Spanish and Quechua spoken in Arequipa.
Quipu : the many-knotted 'cat-of-nine-tails' of the Inca record keepers and

storemen - their 'spreadsheet', and not yet fully understood to this day.
Raymi : Festival
Recreo : large restaurant, specialising in family occasions and group gatherings.
Sanguchería : sandwich bar.
Sauco : Elderberry
Tallarin : pasta.
Tela de cintura : backstrap weaving.
Wachuma : visionary drink made from local cactus in the highlands of Peru (in Quechua *huma* = mind). Named 'San Pedro' in *Castellano*, as St Peter has the keys to the kingdom of heaven. The *Ayahuasca* of the mountains.
Yungas : name for the forested strip on the eastern side of the Andes (*Yunka* = 'warm' in both Quechua and Aymara.
Yupana : the 'table calculator/abacus' of the Incas – an historical marvel that still perplexes and confounds mathematicians.

Cathedral, Cajamarca

A Pinch of Quechua & Aymara

"The limits of my language are the limits of my world"

Quechua: One of the three main languages of Peru, and formerly the principal language of the Inca empire. Today it is still spoken by perhaps 14 million people in South Amerca – the most widely used indigenous language on the continent – mainly in the Andean regions of Peru, Bolivia and Ecuador. However, it is also to be found in Argentina, northern Chile and southern Colombia. As a result of this very wide geographical spread, many individual dialects have arisen, to the extent that the Quechua speaker from the north of Peru may not be comprehensible to the speaker from, say, Andahuaylas. Nevertheless, a few simple words of *Runa Simi* – the language of the people – are offered here for visitors to use and enjoy on their travels.

Hello!	*Imaynallam kachkanki*
Goodbye	*Tupananchiskama*
How are you?	*Allillanchu?*
What is your name?	*Imam sutiyki?*
My name is ...	*Sutiyqa...*
Please	*Ama qina kaychu*
Condor	*Kuntur*
Dog	*Allq'o*
One	*Huk*
Two	*Iskay*
Three	*Kimsa*
Four	*Tawa*
Five	*Pishqa*
Ten	*Chunka*
One hundred	*Pachak*
One thousand	*Waranqa*
Friend	*Wawqey*
Sky	*Hanan pacha*
Water	*Yaku*
Star	*Q'oyllur*
Night	*Tuta*
Rain	*Para*
Fish	*Challwa*
Yes	*Ari*
No	*Mana*
Yesterday	*Qayna punchau*
Tomorrow	*Paqarin*
Mother Earth	*Pacha Mama*
I don't understand	*Manam yachakunichu rimasqaykita*
Sun	*Inti*

Aymara: Not encountered so widely as Quechua but nevertheless spoken by up to 2 million people living on the Andean *altiplano*, principally those in Bolivia, but one fifth in Peru and a small minority in Chile. During the era 400BC to 900AD Aymara was the official language of the Tiahuanaco civilisation. Once again, here are a few words and simple phrases.

Hello!	*Laphi*
Goodbye	*Jakisiñkama*
How are you?	*Kamisaraki?*
What is your name?	*Kunas sutimaj*
My name is ...	*Sutijara ...*
Please	*Mira*
Condor	*Maliku*

Dog	*Anu*
Friend	*Masi*
Sky	*Lakhampu*
Water	*Uma*
Star	*Warawara*
Night	*Aruma*
Rain	*Jallu*
Fish	*Chaulla*
One	*Maya*
Two	*Paya*
Three	*Quimsa*
Four	*Pusi*

Five	*Pesca*
Ten	*Tunca*
One hundred	*Pataca*
One thousand	*Waranka*
Yes	*Jisa*
No	*Janiwa*
Yesterday	*Masuru*
Tomorrow	*Qharuru*
Thanks	*Yuspagara*
I don't understand	*Janiwa entindkti*
Sun	*Inti*

▲▲▲▲

"Evita molestar a las llamas"

Index - Gateways to the Andes

D

E

F

G

H

I

J

K

L

M

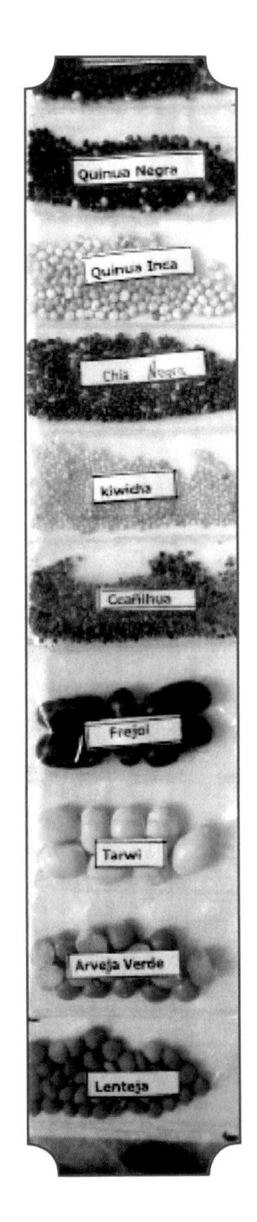
Quinua Negra
Quinua Inca
Chia Negra
kiwicha
Ccañihua
Frejol
Tarwi
Arveja Verde
Lenteja

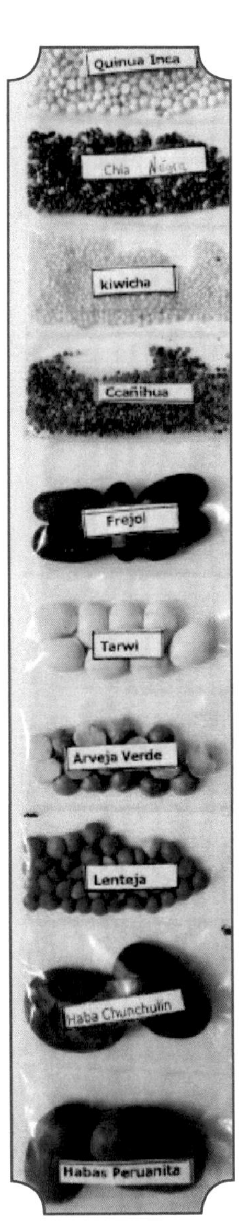
Quinua Inca
Chia Negra
kiwicha
Ccañihua
Frejol
Tarwi
Arveja Verde
Lenteja
Haba Chunchulin
Habas Peruanita

Arveja Verde
Lenteja
Haba Chunchulin
Habas Peruanita
Haba Verde
Maiz Pescorunto
Maiz Checche
Trigo
Cebada

IQUITOS
GATEWAY TO AMAZONIA
AND
PATHWAY TO AYAHUASCA
John Lane
Sixth Edition
The Alternative Travel Guide
ZALZALA
Diary of a Disaster
2005
THE CHRISTMAS PARADE
John Lane
illustrated by Georgina Moir
www.johnlanebooks.com
HOMEWARD BLOWS THE WIND
John Lane
IQUITOS
GATEWAY TO AMAZONIA
John Lane
Fifth Edition
The Alternative Travel Guide
BALTHAZAR and his BENDY BUS
CITY BUS
John Lane
illustrated by Adebanji Alade
EPITAPHS TO DIE FOR
THE POETRY OF BENCHMARKS
JOHN LANE'S
GOOD GRIEF POCKET BOOK
VOL.1
FIRST STEPS in the ENCHANTED FOREST of stitches
by Bella & John Lane
Illustrated by
PERUVIAN PRACTICE
TRAVELS WITH LA SEÑORA
"A fine comic writer"
MIRANDA FRANCE, DAILY TELEGRAPH
JOHN LANE
CHICKEN STREET
AFGHANISTAN BEFORE THE TALIBAN:
CLEARING THE DEADLY REMNANTS OF WAR
JOHN LANE
Foreword by Martin Bell
THE PRINCESS AND THE PIGEON
John Lane
Illustrated by
IQUITOS
GATEWAY TO AMAZONIA
by
John Lane